Designing Apps for Success

Designing Apps for Success

Developing Consistent App Design Practices

Matthew David and Chris Murman

Focal Press
Taylor & Francis Group

NEW YORK AND LONDON

First published 2014
by Focal Press
70 Blanchard Road, Suite 402, Burlington, MA 01803

and by Focal Press
2 Park Square, Milton Park, Abingdon, Oxon OX14 4RN

Focal Press is an imprint of the Taylor & Francis Group, an informa business

© 2014 Taylor & Francis

Notices

Knowledge and best practice in this field are constantly changing. As new research and experience broaden our understanding, changes in research methods, professional practices, or medical treatment may become necessary.

Practitioners and researchers must always rely on their own experience and knowledge in evaluating and using any information, methods, compounds, or experiments described herein. In using such information or methods they should be mindful of their own safety and the safety of others, including parties for whom they have a professional responsibility.

Product or corporate names may be trademarks or registered trademarks, and are used only for identification and explanation without intent to infringe.

Library of Congress Cataloging-in-Publication Data

David, Matthew, 1971–
  Designing apps for success : developing consistent app design practices / Matthew David, Chris Murman.
    pages cm
  Includes bibliographical references and index.
1. Mobile computing.   2. Smartphones—Programming.   3. Application software—Development.   I. Title.
  QA76.59.D3875 2014
  005.36—dc23
  2013040550

ISBN: 978-0-415-83441-4 (pbk)
ISBN: 978-0-203-50587-8 (ebk)

Typeset in Helvetica Neue
By Apex CoVantage, LLC

Printed in the USA

SUSTAINABLE
FORESTRY
INITIATIVE

Certified Chain of Custody
At Least 30% Certified Forest Content
www.sfiprogram.org
SFI-01268

SFI label applies to the text stock

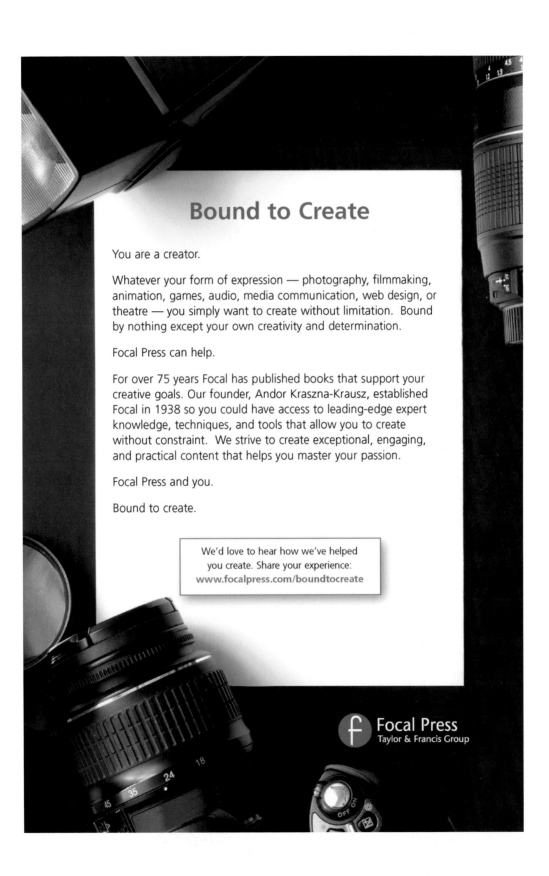

Bound to Create

You are a creator.

Whatever your form of expression — photography, filmmaking, animation, games, audio, media communication, web design, or theatre — you simply want to create without limitation. Bound by nothing except your own creativity and determination.

Focal Press can help.

For over 75 years Focal has published books that support your creative goals. Our founder, Andor Kraszna-Krausz, established Focal in 1938 so you could have access to leading-edge expert knowledge, techniques, and tools that allow you to create without constraint. We strive to create exceptional, engaging, and practical content that helps you master your passion.

Focal Press and you.

Bound to create.

We'd love to hear how we've helped you create. Share your experience:
www.focalpress.com/boundtocreate

Focal Press
Taylor & Francis Group

DEDICATION

For my wife and kids. Love you all.

Contents

Preface

Mobility is no longer a fad. The reach of mobile technology is greater than that of any other technology. This was seen clearly when the popular mobile game Temple Run 2 was released: it hit 6 million downloads in 24 hours.

It is often said that Apple launched the new mobile ecosystem with the launch of the iPhone. It didn't. It was the launch of the iTunes App Store that changed the world. Apple introduced a way to download an app with one click and have it running on your smartphone. Apple made installing software as easy as buying a song or changing the channel on your TV. The concept changed how people installed software. So far, Apple has seen more than 50 billion apps downloaded.

As was expected, competitors quickly came to market with their own smartphone products. The leader is now Google, with its Android operating system. Today, close to 80 percent of all smartphones sold worldwide run Android. That is an impressive number. What is more impressive is that Google has now activated more than 1 billion Android devices. To put that in context, the number of Android phone being activated by Google is now three times the number of personal computers sold. If you now include the number of devices running Apple's mobile operating system (iOS), you have 1.6 billion devices; that is more than the total number of computers being used worldwide.

The launch of the smartphone is only phase one of the mobile revolution. Phase two came with the launch of the iPad, in 2010. Tablets immediately caught on in businesses. The price of iPads, however, kept them out of the hands of most consumers. This changed when first Amazon with the Kindle Fire and then Google with the Nexus 7 released seven-inch tablets selling for less than $200. Today, you can buy a tablet for as low as $159.

Mobility continues to evolve. Wearable devices, such as the Samsung Gear smartwatch and Google's Glass, clearly show that smart devices can become fashion accessories. The auto industry is also engaging with mobile devices. Smart vehicles, such as GM's OnStar system, now come loaded with apps that run from the in-car console.

Cisco believes that there will be close to 10 billion mobile devices by the end of the 2010s. Each person will possess more than one smart device. Consider the potential reach.

The focus of this book is to outline how you can design solutions that can be published to this unique global community. The book focuses on three key areas:

- Getting started with your ideas and your team
- The technologies you can use to build apps
- How to promote your apps

Each section is broken down further into chapters that zero in on a specific topic. Every concept and every technology in this book has been tested and is being used by my team and clients today. Currently I am chief digital strategist for Compuware, where I lead the direction of our Mobile Center of Excellence. Before taking this job, I led marketing and evangelism for a mobile startup that has gone on to win awards across Europe. And it all started when Apple introduced "the app."

Acknowledgments

This is my fifth book for Focal Press. It is not easy to write a book. Yes, I know, writers will tell you that it takes just one word at a time, but it is a lot of words. Creating a book is not something done in a vacuum. You have friends, family, and coworkers who support and help you all the way. The team at Focal Press, David and Carlin, are rock stars. This book is as much their book as mine (really, we should have their names on the front covers).

The next person I need to thank is the co-author, Chris. Dude! I needed help and you stepped and did a mighty fine job. I'll work with you any day of the week.

The material I cover in this book is from the work I have done over the past few years in the mobile tech world. There are some people I do need to make a big shout-out to. These folks changed my world and how I see it: Andy McCartney and the founding team at JamPot (guys, you are visionary); the folks at MillerCoors in Milwaukee (thanks for the beer, too!); and my brilliant team at Compuware—the teams on the fifth and eighth floors are making magical solutions happen!

Finally, I really need to thank all of the people on Twitter and Google+ who are quick to answer my questions. Seriously, you guys rock!

SECTION

1

Designing Apps to Work

Putting Apps to Work

Welcome to a paradigm shift. The reset button has been pressed. You have chosen the red pill and will follow Alice down the rabbit hole.

The mobile revolution is now part of our lives. Look at how mobility drives what you do all day: your phone is a now a computer in your pocket. Your tablet is a second screen when you are watching TV, and your car now has a centralized infotainment system. We now engage with technology all the waking day—it is with us wherever we are, as shown by the increasing choice of mobile devices as shown in figure 1.1.

The boiling point for the dramatic change in how technology is part of our lives was not reached because of the iPhone. There have been other smartphones on the market. The boiling point was reached when Apple introduced the App Store, a new way to consume content.

Figure 1.1: The choice of mobile devices includes laptops, smartphones, and tablets.

THE RISE OF THE SECOND SCREEN

Why is mobile part of your business plan? Let the numbers speak for themselves:

54% of all US Americans now have a smartphone (IDC).

25% of all Web traffic now comes from a mobile device (Cisco).

Worldwide, smartphones now outsell feature phones (Gartner).

What is the meaning of these numbers? You have a mobile device, your customers have mobile devices, and your business partners have mobile devices. Now is the time to start using them for business.

Figure 1.2: The second screen experience.

My three-year-old daughter was using apps years before she knew how to use a computer. Why was it so transformative? Because anyone can install an app.

App usage is successful for three key reasons:

- One-click install—you don't need a lawyer to read the legalese anymore
- Natural user interface through touch
- Limited screen space, forcing developers to zero in on key functionality

In this chapter you are going to be introduced to the foundational elements of app usage in your organization. The chapter has been split into the following sections:

- Defining How Mobility Drives Additional Business
- Just Get Started

- Creating Your Mobile Strategy
- Plan to Measure Success

The goal of this chapter is to get you started thinking about a mobile strategy and how you can use mobility in your business. Using iPhones, tablets, and apps will give you the knowledge you will need to step through to the next phases of mobile solutions.

Figure 1.3: iPads being used in hospitals.

MOBILE IN YOUR BUSINESS

Mobile is now in your business. How do you know that? Balance these numbers against your organization:

94% of Fortune 500 companies have active iPad programs.

25% of your customers arrive at your website on a smartphone.

Almost every adult American who has a phone has a smartphone. Indeed, the smartphone category is so pervasive that we are now simply calling this category "phones."

Defining How Mobility Drives Additional Business

Have you heard the joke about the consultant who comes into your business and tells you that you need to be a mobile-first organization? It's a good one: the consultant will tell you that you need to redesign all of your tools to run off an iPhone.

But the consultant won't tell you how you are going to make money, improve internal processes, or drive value to your strategic partners.

So let me be clear: we are *not* a mobile-first world. We are and always will be a *business-first* world.

Mobile is an exciting and engaging technology, but it is simply a technology, and technology does not drive business. To use Forrester's POST definition, this is how we drive business:

P—People

O—Objectives

S—Strategy

T—Technology

It is no coincidence that technology is the final part of the puzzle. So, let's look at different types of businesses, their objectives, and how they are engaging with their clients and understand how technology, mobile in this case, helps them.

MOBILITY FOR THE SOLE OWNER

I have been a small-business owner for years. I love it for the same reason that millions of people around the world love being a small-business owner: control. At the end of the day I am the boss, the CEO, and the cleaning crew. Do I care that I have so many roles? Not one bit!

There are many different types of sole owner. Many started like I did: with a paying hobby. You found something that you loved doing; others liked it too and were willing to pay you something to do it for them. Your hobby is now a new source of revenue, and you are the boss of a brand-new company.

Figure 1.4: Leading smartphones including Android, iOS, and Windows 8 devices.

THE PRICE OF ENGAGEMENT

It does not cost much to get involved with mobile. Consider these numbers:

- Top-of-the-line smartphone—$250–$400
- Last season's smartphone model (6–12 months old)—$49–$99
- Tablet (10-inch model)—$499–$999
- Tablet (7-inch model)—$99–$329
- Apps—FREE–$30

As you can see, the apps have the lowest price point. Most of the really good apps have a really generous "try before you buy" plan with built-in upgrade paths. Check out Evernote and Google Docs.

Most smartphones charge a premium for today's top design, but yesterday's models are now so good that the argument for having the latest and greatest is getting weaker by the day.

I have almost a dozen different tablets and have used every version of the 9.7-inch iPad. So why do I always have my Nexus 7 with me? Well, first, it is really cheap ($199), and I don't mind if it gets broken. Second, it fits perfectly in my jeans back pocket, so I don't have to carry it everywhere.

The bottom line is that you have great choice at affordable prices when it comes to selecting apps, smartphones, and tablets. The good news is that all of your options are excellent.

Figure 1.5: Evernote running on an iPad.

While you are not a Fortune 500 company, defining your business strategy should not be any different from the practices of a large company. Actually, you have a distinct advantage over the very large companies: your operating margins are very low. In my case, I love to write and teach about the technologies and solutions I put in place. All I needed to get started was a computer. Fortunately for me, I had a boss who was throwing out a five-year-old desktop computer and who took pity on me and gave the computer to me for free. With my new free (but really slow and generally junky) computer, I wrote my first book. My objective was to write the book; the strategy was to make money from the book sales and to invest in new computer technology.

Figure 1.6: Google Docs running on an entry-level Android tablet.

Today I still write. The people part of the equation is still the same. The objective has not changed.

But I will tell you what has changed: the strategy and the technology.

My new book is being written on a MacBook Pro. I take it everywhere (currently I am sitting in the kitchen writing, but yesterday I was writing in Madison, Wisconsin, and next week I will head to LA). What I call my "big writing"—the work that takes me a while to complete and requires me to focus for a few hours—I do on my laptop.

But this is where technology is changing my strategy to reach my objective. I save all of my files to Dropbox. Dropbox is a file-sharing cloud service. It is a brilliant tool for me for the simple reason that it is free (you have to love free stuff when you are the sole owner). But this is where the strategy piece comes into play. To get a book written you have to have several people involved:

- The writer—that's me.
- The technical editor—Ryan is filling that role for this book.
- The editor—not sure who that is, but he or she normally sends me back my chapters covered in red ink.
- The project coordinator—Carlin is leading this role (she will identify the layout team and the editor).
- The acquisitions editor—The Man with the Plan, the guy who makes it all work, is David.

I run my business, but I am dependent on a lot of people to get the work done. Ten years ago I would complete a chapter and e-mail it to my project coordinator. That person would then send the files on to my technical editor and my editor. They would both edit and send the files back to me. I would then have to consolidate edits from two new documents.

I would make the changes (so I would now be on version four of the document) and send the chapters back out to the project coordinator. This would happen a few times. Finally, on version 10 or 11 of the chapter (and, normally, I would lose an e-mail here and there and miss a version that needs to be edited), the files would get sent over to the layout team.

The strategy of engaging all of these people is still the same. Technology is now making it run more efficiently. With Dropbox I can now share all of my files with everyone on the team. As soon as I edit a document they will see the file on their desktops. There is no need to e-mail the document to anyone. They have it and can now edit it. What is even better is that the editor can send me a text, short e-mail, or Google Chat message telling me that he or she has made the edits for my review. But I don't review edits on my laptop. No need. Most edits are calling for clarification on a statement and do not require hours of time spent tapping away on a keyboard. I make the edits on my smartphone (currently I am carrying an iPhone, but that changes) or my iPad or Nexus 7.

Figure 1.7: Dropbox syncing images on an iPhone and an iPad.

What has changed in the past decade is that technology is now contextually applied to where I am at that point in time. This is powerful for a small-business owner.

It is clear that mobility is a massive part of your business—not because it is cool (well, maybe a little) but because we as humans are mobile. We don't stay in one place, and it drives us all nuts when we do. Your technology can now come with you.

The most important element of mobility to a sole owner is the price of engagement. When I started, I needed to buy a computer, a $1,500 investment. I was almost prevented from starting my writing career because of the price. Today's smartphones are incredibly cheap and very powerful. Tablets can be purchased for as little as $99. Think of that for a moment—that is the same price as a tank of gas!

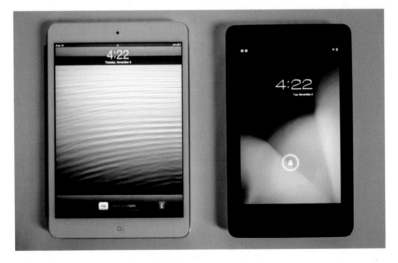

Figure 1.8: A Nexus 7 and an iPad Mini next to each other.

Once you have your phone or tablet, the cost of adding powerful communication tools such as e-mail, file sharing, and document creation is nothing or nearly nothing.

In other words, what would have cost me an investment of close to $3,000 ten years ago I can get for $400 today for an iPhone and a Nexus 7 tablet and then an additional $30 to $50 for software. Incidentally, the apps are really easy to install, and I do not need to keep an IT person on staff.

To sum it all up: as a sole owner you need to work with partners to get your work done. Your objectives are still the same. But, technology, specifically mobile technology, is dramatically changing the effectiveness of your communication and processes. And the cherry on the top: the price is radically lower than ever before.

SMALL-BUSINESS MOBILITY BUSINESS OPPORTUNITIES

In many ways, the job of a small-business owner is not much different from that of a sole owner. The challenge is scale. You have more people, more work, and larger amounts of money moving around.

How can mobility help? In exactly the same way: it can help you focus on your business, understand where your people are located, and leverage technology for where they are. For instance:

- Realty agents—you're away from your house but still need to check e-mail, post photos of a new property, and schedule times to meet with prospects. Do it all from your phone! It is the perfect tool for you.

- Restaurant owners—move your menu onto cheap 7-inch tablets and have your patrons tap the photos to order the meals. This allows your staff to focus on their roles of creating a perfect eating experience.

- Package delivery managers—use GPS on a phone to find clients' locations and then have the clients sign for their packages on a smartphone.

WHICH DEVICE SHOULD YOUR COMPANY CHOOSE?

Good Technologies published a survey outlining who is using what devices within corporations.

- 77% of all devices are either iPhone or iPad.

- 22.7% are Android (but the number is rising year over year).

- BlackBerry and Windows barely register.

Why does Good Technologies know this information? It is one of the leaders of mobile device management solutions (along with AirWatch, MaaS360, and MobileIron) and has millions of customers using its tools.

What is really interesting is that one-third of Apple's activations on Good Technologies' network are for iPads.

Figure 1.9: Mobile device management seeks to address three key things about your mobile assets: enforcing policy, knowing where you asset is, and securing data on the device.

The focus is to use technology for where you are connecting with your customer. Think of it like this: while you are talking with a customer you are engaging in a dialogue. Over the past 20 years we have become used to making statements like "OK, I will send you the documents later in an e-mail." That one statement breaks the dialogue.

Try this: you are talking with a client and showing her a contract, and you say, "I am now sharing the contract with you. Check your phone." The client takes out her phone (don't worry, your client does now have a smartphone), and she has the document on her phone with a place where she can sign. The client signs the document. The contract is legal and is completed in minutes, and you both have copies of the documents—signed, sealed, and delivered.

LARGE-SCALE MOBILITY SOLUTIONS FOR ENTERPRISES

Enterprises have many challenges. Employee count, size of projects, geographical and cultural diversity, many different technologies . . . the list goes on and on.

But it can be argued that every enterprise is looking to complete three objectives:

- Deliver customer value (through products and/or services)
- Improve employee efficiencies
- Speed up throughput in the partner channels

Again, technology has not changed these three elements. The process is the same, and the end deliverable is the same. What has changed is velocity.

THE RISE OF THE SECOND SCREEN

Many of us still have a primary screen for work or pleasure. It's your laptop or your TV. We spend hours in front of both. But the rise of the smartphone and the tablet is giving an opportunity for a second screen. Think about this: do you want watch TV with your mobile device now in your hand? Bet you do.

Consider a second screen in your work. Send tasks that require only viewing or simple responses to a second screen.

Figure 1.10: Fresh fruit quality is a top concern for Dole.

Here is an example that Dole Fresh Fruit Company put into place. Dole's business is based on selling bananas. The challenge is simple: bananas are a commodity, and profits are razor thin. So how can you make the business profitable? Go into the grocery section of a store and buy some bananas. The prices for different brands are likely to be similar. The differentiator is quality. Do you buy the bananas that are going brown or the fruit that is yellow and fresh?

Turns out, we almost always choose quality fruit.

To deliver the best fruit to the store, Dole has a small army of fruit inspectors who assess the product when it arrives at the ports and docking stations. The inspectors assess the quality of the fruit, make notes, take photos, and send the information to the company HQ, where the report can be assessed.

Here is the problem. It takes two weeks for Dole to make a decision on any fruit product. The reason: inspectors are moving all day long from one location to another and make notes on a clipboard as they go along. At the end of the day (or sometimes a day or two later), the notes are consolidated into an Excel spreadsheet and sent via e-mail to HQ. When they arrive at HQ, all of the reports are consolidated into one. This takes another day or two. The final report is sent out to the team that assesses the fruit report. This can take another day or two.

The whole process takes 14 days. But what if a fungus is reported? The current process does not allow for anyone to be duly notified for as much as 14 days.

The opportunity here is this: the people doing the work know what they are doing (all the way up and down the line); the process is the right process (inspectors need to be in the field to report on fruit and the business leaders need to see the data from all of their inspectors); what needed to be changed is the technology.

Mobile phones, it turns out, are perfect. Dole created an app that does the following:

- Capture key data points from the fruit being inspected
- Take a photo of the fruit
- Allow the user to hit a submit button and send the data back to corporate HQ
- Automatically consolidate the data

Using mobility, Dole has reduced the time from when the inspectors examine the fruit to when the business leaders can make a decision from ten days to one day. That is an improvement.

KEEPING UP TO DATE ON INFORMATION

My top source for sharing knowledge with people doing the work of mobile enterprise engagement is to speak directly to my peers. The best group for this is on LinkedIn: Enterprise Mobility Group. Join the group today. You will be glad you did.

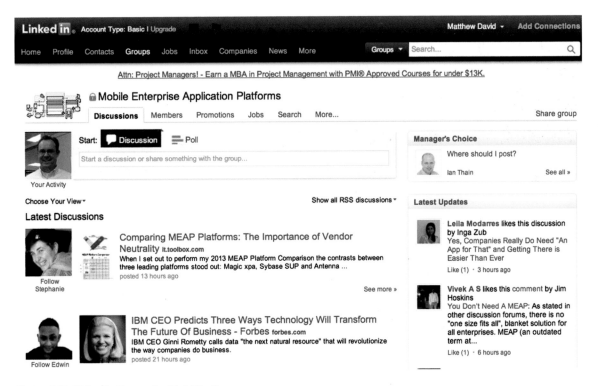

Figure 1.11: LinkedIn Enterprise Mobility Group.

The improvement in the app was realized in less than 90 days.

Additionally, Dole is now delivering its training manuals through the app. The traditional approach to delivering the training manual was to send out, once per year, a large three-ring binder that would sit on the shelf of the fruit inspector's cubicle. But the fruit inspector is seldom in the cubicle. The manual is now part of the inspector's app. The benefit is that the manual is now with the inspector all the time, and updates to the manual can be pushed out several times per year.

This is one example where enterprises are looking at mobility as a way to move technology to where the user is working. Don't force your employees or customers to be tethered to a desktop computer. Move the information to where they are right now.

Just Get Started

Fail, and fail fast. It is an interesting expression. It is very applicable to mobility. So what is the baseline for this expression? Why do we want to fail fast? Well, let's look at this concept: in baseball, when you come up to the plate, you are considered a superstar if you hit the ball one time in every three at-bats. This means that to be considered a superstar hitter, you miss the ball most of the time. You fail. But you pull your arm back and gear up for the next throw. Why? Because you have learned something about the pitcher. You now know how he will send the ball to you.

Yes, the hitter fails. But failing fast and learning and moving on to something that works allows him to be successful even faster.

Don't be afraid to fail. Be prepared to learn. And then to succeed.

Your mobile strategy will go beyond the iPhone and iPad

Figure 1.12: Your mobile strategy is not an iPhone or iPad; it is your business.

This can be said of your mobile strategy. Many companies are struggling with where to start with mobile. Frequently I hear the comment "I don't even know where to start."

Here's the secret: no one knows where to start. Mobility is not like previous technologies. It is changing and morphing very quickly. Yes, we have smartphones, but we also have tablets (and I'd argue that 7-inch tablets are different from 10+-inch tablets) and in-car systems, and what about wearable technology such as smartwatches, wristbands, and glasses? Yes, this field is growing, adapting, and changing our lifestyles much faster than we ever thought possible.

So here's what you do: do something. You already have a smartphone and a laptop. Begin to use cloud tools such as Google Docs and Evernote to share data so that you always have your information with you. Try new tools until you find the tool that sticks. For me, that one tool that made me realize everything has changed is Dropbox: it is free and allows me to share my documents with my team. I like it so much that I now pay for the higher-level product.

MOBILIZE YOUR BUSINESS

The first step you want to take in mobilizing your business is easy: get e-mail, calendar, and contacts onto your phone.

Next, find something that will improve your business process. Both Google and Microsoft are making their office productivity tools available to smartphones, tablets, and traditional computers.

USING GOOGLE DOCS FORMS TO EXPERIMENT WITH MOBILE DATA COLLECTION

For Google Docs users, leverage the Form tool. The tool allows you to create forms that can be sent via e-mail, embedded into Web pages, or updated on phones and tablets. The data from each form that is completed are stored in a spreadsheet. Data collection solutions such as Forms can carry a high return on investment—they take only minutes to create, and measuring success is easy.

Microsoft's SharePoint now fully supports multiple devices. SharePoint gives you several key solutions right out of the box:

- Document sharing

- Content management

- Workflow

Documents can be viewed and read from your iPad in SharePoint. This allows teams to work quickly from a central document instead of sending many documents through e-mail.

Forms and content are key elements that have been part of SharePoint since the first release. Now your teams can edit and add new content from where they are working.

Figure 1.13: SharePoint 2013 running on smartphones and tablets.

SharePoint's third strength is workflow. Built into SharePoint is a sophisticated workflow engine. You can create forms that require sign-off or additional information using InfoPath. What is interesting, however, is that the data and workflow are separate from the presentation layer. This means that data can enter SharePoint and be moved through SharePoint, but the interface layer can be a custom app. For instance, an expense report can be completed on a laptop computer, with the approval request sent to a form on an iPhone.

GETTING STARTED WITH COMMERCIAL APPS

The highest risk you can experience with mobile apps applies to those that face the public. Once your app is out there, it had better work. If it does not, then you will experience the wrath of the Internet—Twitter, Facebook, and Google+ are great places to vent your anger on a bad app.

There are three companies that have done a great job with their commercial mobile apps: Starbucks, Walgreens, and Tesco/Home Plus.

For me, the Starbucks app is successful because it follows the mantra of "less is more." The goal of the Starbucks app is based on three concepts:

- Let you pay for your purchase from your phone

- Lets you download the free "song of the week"

- Helps you find the nearest Starbucks store

Without doubt, the top feature is the ability to purchase tea and coffee with your phone. The Starbucks app gives you an easy-to-use tool that lets you upload your gift cards onto your phone. The bar code from the gift card is now on the phone.

OK, that is cool, but what I really like is this: when I make a purchase, my card will deduct the balance in real time. There is no need for me to guess how much of a balance I have left on the card. I am told my balance while I am waiting for my latte to get steamed.

Figure 1.14: Paying for coffee with a Starbucks app.

Starbucks is also leveraging "gamification" in their app. "Gamification" is a term that has been running around for a while. It means transforming the tasks we do and making them fun. Turns out that smartphones are great devices for gamification. We like little, short games like a level of Angry Birds or a quick run down the path on Temple Run. With this in mind, Starbucks rewards me on my iPhone app each and every time I use the app to make a purchase. I get a little gold star that bounces around a virtual coffee cup. Yes, I can hear you now, "Very cute, but what is the value?" The value is tied directly to rewards. The more stars I get, the more free cups of coffee I will receive. Yes, bring on those little stars.

Walgreens also sees its app as critical to its business. The success of the Walgreens app is proved by its always being among the top ten apps in app stores. This is driven by Walgreens' focusing on three things in its app:

- Why do our customers need another app?
- How can the Walgreens app benefit our customers?
- How can Walgreens reward customers with the app?

Walgreens is a Fortune 50 company with a massive chain of stores with a specific focus on pharmaceutical supplies. But there are other stores and other places I can get pain medication. Why Walgreens?

Turns out that convenience trumps price. As a Walgreens customer, I see Walgreens doing just one thing for me: fulfill my prescription. At one time I had to walk into the store with a piece of paper

Figure 1.15: The Walgreens app lets you renew a prescription, print photos, and view the latest deals in the customer loyalty program.

with the doctor's scribble on to get my medication; later I could call in my prescription. Today, with the Walgreens app, I can scan the bar code and select when I want to pick up the next prescription. The effort for me to fulfill my prescription has gone from taking time out of my day to stop into the store, to spending ten minutes on the phone, down to a zap on my iPhone. The process is the same, but the speed and convenience are greatly improved.

You know what else Walgreens is really good at: processing photos. Yes, we live in a digital world, but we still like to have printed copies of really good pictures. I like to have Walgreens print my annual Christmas cards with the kids' picture on them. The Walgreens app also allows me to upload photos.

When I get to the Walgreens store I can now pick up my prescription and my photos. A third tool Walgreens has added to its app is the customer loyalty program. Bam! Discount!

With one app Walgreens has been able to consolidate the following:

- Prescription requests
- Printed copies of digital photos
- Discounts I did not even know about

The whole process is clean and simple.

WALGREENS REACHES OUT TO DEVELOPERS

Walgreens is extending the functionality of two of its core features out to developers: prescription refills and photo printing. If you include the features in your app, you will get paid each time someone uploads a photo!

Leveraging an iPhone/Android SDK and RESTful APIs, it takes only about four hours to add the features into your app.

The third example of a store that is leveraging mobility as a tool to deliver products through a new channel is: Tesco/Home Plus.

South Korea has a problem: it is a peninsula country with limited land for its growing tech-savvy population. Grocery shopping on Saturday can takes hours and is not much fun.

Where there is a problem, there is someone looking for an opportunity. Welcome Tesco, one of the world's largest grocery store chains (yes, Walmart, Tesco is looking to get bigger than you!). Tesco wanted to be part of the shopping experience but was ranking in a low #5 spot for customers in South Korea when it comes to grocery shopping. To address this issue, Tesco rebranded as Home Plus and moved the shopping experience.

Today, when you are waiting for your next train in the subways of South Korea, you will see billboards lined with products like a virtual shopping mall. Tesco/Home Plus took photos of milk, eggs, cheese, and all the items you buy in the grocery store, along with QR codes, and is lining the subways with the images. With your phone you can scan the QR code, add the item to your shopping list, pay for your groceries, and schedule when you want the items delivered while you are waiting for your train. Suddenly your afternoon is not going to be spent fighting your way through a grocery store. Tesco/Home Plus has given back the gift of time to its customers.

Tesco/Home Plus is now the #2 grocery chain in South Korea. It did that without opening a new store.

Figure 1.16: Tesco/Home Plus virtual stores use mobile phones to take orders and schedule when your groceries will be delivered.

These three companies are examples of businesses that have addressed a specific problem with their customers by leveraging smartphones as the tool to resolve the problem.

Creating Your Mobile Strategy

When a new media channel is available, it is worth taking time to assess the value of the channel and how it applies to your business. In other words, define your strategy.

MOBILITY FOR THE SOLE OWNER

You have the direct relationship with your customers. You manage the phone when they call. You hand them the receipt after they make a purchase.

You do it all.

So don't do it at home on your laptop, where your customer is not located. Do the work and take the payment where your customer is.

Essential tools should be the following:

- Use Square to complete your transactions with major credit and debit cards on your phone.

- Use PayPal on your phone as another way to take payments.

- Manage your money with Mint.

- Share your documents with Dropbox.

- Set up virtual communities with your clients using Google+.

- Share photos of the work you are doing on Pinterest.

As the sole owner of the company you have the greatest return on investment for your investment in mobility. A smartphone and a tablet are all you need. There are hundreds of apps you can leverage that will make your business better. My recommendation is to start using them.

Figure 1.17: Square allows you to take any major credit or debit card.

SMALL-BUSINESS MOBILITY OPPORTUNITIES

A small business has a focus on one or two specialty products or services. You may insure jewelry, change the oil in your customers' cars, or run a law office. What you are all doing is interacting with technology.

Look at the money you spend on technology: not just the hardware such as laptops but also the service time of technicians.

Now, how would you like to reduce that cost by a factor of ten? This is where mobile and the cloud will help you. Services such as Office 365 mitigate the need to have onsite e-mail, the hardware to support your e-mail, and the technician to run it. Cloud-based office solutions are also wired to run on your phone and tablets.

Your immediate strategy should be to offload new services to a cloud solution with access from your existing laptop and new smartphone. The goal is not to replace your existing tools but to offload task-driven actions (such as checking e-mail, invoicing, and action requests) to your phone.

INCREMENTALLY EXCHANGING LAPTOP SERVICES WITH MOBILE

Take one repeating task—such as scheduling meetings—and move the action to your phone. What you will find is that you can schedule meetings while standing in the elevator, allowing you to more efficiently use your time. Do this for one month with your team. Next month, choose another task and move it off your laptop. What you are creating is a more efficient organization.

LARGE-SCALE MOBILITY SOLUTIONS FOR ENTERPRISES

Are large enterprises different from small companies? No. The only difference is scale. Companies with 5,000, 10,000, or 20,000+ employees have a challenge driven by scale. To address the issue of scale I recommend consistent interactions of smaller projects. Not because smaller can operate under the radar, but the opposite. It is easier to measure smaller projects. The end result is to present successful solutions that have clear and defined goals.

There are five key steps to enterprise mobile maturity that you can use as milestones to present to your organization. The five steps are:

1. Decide on a mobile strategy—put in place the leaders who will drive success.

2. Implement device management—select the tools that will give your mobile devices the best functionality.

3. Leverage third-party apps—allow your staff to become comfortable with mobility by implementing support for apps such as Concur for expense management and Evernote for meeting notes. If you have SAP, use the many SAP apps available.

4. Build your first app—understand what it takes to build a mobile app by building a mobile app. Choose a process that is small and can be measured.

5. Define your mobile strategy—understand what you need to do as an organization to leverage technology to improve revenue, process, and partner communication.

The goal of your strategy is to present a clear vision for how mobility will become part of your company's toolset.

SAP'S MOBILE ENTERPRISE APPLICATION PLATFORMS

Do you run SAP? Over the past couple of years, SAP has been investing heavily in mobility to support your distributed workforce. Thanks to tools such as Afaria for device management, Business Objects for mobile business intelligence, and Syclo for rapid app development, you now have a suite of tools you can leverage to deliver mature mobile solutions.

Plan to Measure Success

How do you know you have been successful? Is it is a hunch? A feeling, maybe? How about you build in some metrics? The balance of this book is all about designing, building, and marketing your apps. But, before you build an app, look at what you are doing. Almost always, an app is a new channel to an existing problem. Measure your problem. Understand what the cost is to keep that business line going. Then build your app and measure the success.

You want to do this for a few reasons:

- Many business leaders are skeptical about mobility—numbers dictate success.
- Employees do not like change—delivering a solution that is easy to use mitigates anxiety.
- Money always talks—creating a new revenue channel is expensive, but showing how it can be profitable is very fulfilling.

Now, you have what you need for your plan. Let's get started building.

CHAPTER 2

Designing Your App

App design is very different from desktop design. To begin with, you have a much smaller screen (between 3.5 inches and 5 inches), and the primary controller is a finger. The approach to designing a mobile app, however, is not different. We still use paper and pen to sketch out designs and then move to illustration tools to mock-up designs.

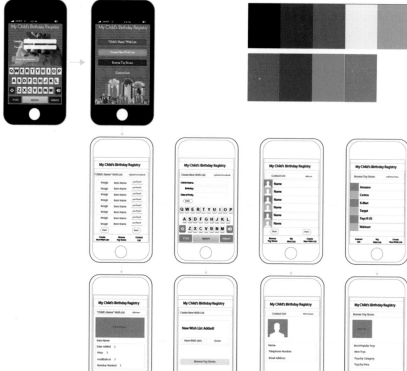

Figure 2.1: Sketches are cheap to create and provide an immediate point of view for your customer, explaining how your app will work.

One focus of this chapter is to draw out the tools you can use to make a mock-up design for mobile apps you are creating. Without doubt, it is much cheaper to go back and forth with a client during the design phase of an app when the app exists only as an illustration. It gets much more expensive to change things when you have hired programmers to create code.

A call to action that was given to me regarding app design: don't design yourself into a corner. At the time I was not clear what that meant. I mean, come on, I am creating apps that sit on an iPhone.

Yep, that was mistake #1. I was designing for just the iPhone. Back in 2010. And I was not thinking about any other type of screen. Duh!

There are now thousands of different mobile operating systems.

As you create your designs, you will need to keep in mind many different factors, such as these:

- Each mobile operating system has its own design guidelines.

- A smartphone interface is *not* the same as a desktop interface. Do you design your solutions to scale, to different screen perspectives?

This is a question I will be addressing, and I hope you will not be faced with the challenge I faced when a client asked if I could take my iPhone design and bring it to the Kindle Fire (yep, different interface, different OS, different size screen—a whole ball of wax of different challenges) and left me scratching my head until I figured out how to complete the task.

Sketching Your App

Do you carry a sketchbook with you when you meet with clients? Or do you outline a solution on a napkin? Sketching is the first step toward creating a solution. It is fast, easy to adapt to changes, and can help you visualize a solution for a client within minutes.

You know what? I am a terrible at multitasking. "One problem at a time" is my motto. It isn't as sexy as having ten screens open all at once on the desk with Twitter, e-mail, and other apps pinging every minute for my attention. But I do get work done.

Leading mobile technologies such as iOS, Android, and Windows 8 have done something I have been clamoring for: gotten rid of windows.

New mobile interfaces are task driven. Pull up a screen and get the work done. To this end, as

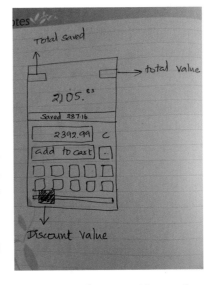

Figure 2.2: Draft out your ideas so the client can see what your vision is.

you sketch out a screen in your new app, place a focus on completing a task. Ask yourself the following questions when sketching a new screen:

- Why do you have the screen?
- What will you get done in less than 30 seconds within that screen?

THE GROCERY LINE RULE

Remember the last time you bought groceries? What did you do while you waited to pay? Check e-mail? Play a level on Angry Birds? Send out a tweet? You completed a task in the time you had at hand—about one to three minutes, while the person in front of you checked out his merchandise. This is called the Grocery Line Rule.

Each screen you create should be task driven, with a result that can be completed according the guidelines set by the Grocery Line Rule. If the task takes more time, then you need to be brutal and get rid of it.

THE VALUE OF SKETCHING

Sketching out your app allows clients to see their ideas and lets you illustrate how an idea can scale from phone to 7-inch tablet to full tablet and to other screens. For many of your customers, this will be their first mobile app. Sketching gives you the ability to both show what can be done with a mobile device and give your client the opportunity to ask questions and run "what if" scenarios.

Fundamentally, your sketches will be the first step in the life of the app you and your team will create.

SKETCHING TOOLS TO USE

The first time that I began working on building out an app for the iPhone, back in 2008 (which now seems like a looooong time ago), there were no easy-to-use tools to illustrate an app. We used a white board and paper. Very low tech, but it really did get the job done.

Today, you have a slew of high-tech tools you can use to sketch out your app for your client. So, let's step through some of the great tools I have used and see if there is a fit with you and your team.

WHITEBOARDS ROCK!

An episode of *House* would not be complete without a trip to the whiteboard. Throw your ideas onto the screen and erase what you don't like. Whiteboards are a great place to rapidly iterate through ideas. And they are bigger than a notepad.

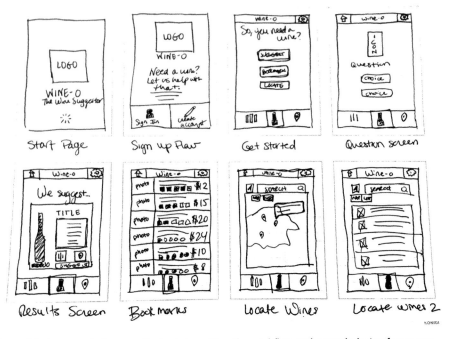

Figure 2.3: A notepad gives you the tools to outline the workflow and a rough design for an app.

TOOL #1: PEN AND PAPER

OK, I am going to harken back to the days when I started as an app developer. My magic tool is still a pen and paper. Yep, a good pen. Even better, a sharp pencil. There are two reasons why I like to use a pen and paper when working with a client:

1. We can all relate to a pen and notepad—clients are not intimidated by the technology and can focus on the subject matter.

2. Paper is cheap and can be readily scrapped if a better idea comes along.

My paper of choice is a moleskin notepad that I can slip into my pocket. The paper size is similar to a 7-inch tablet (how convenient), and it is relatively small in size. The small page size forces the designer and the client to keep their notes and sketches specific. In other words, you are training your client to zero in on the key tasks in the app.

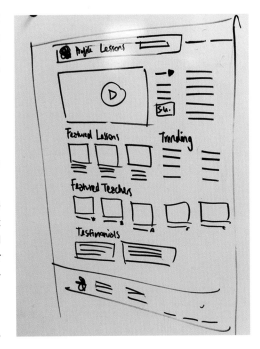

Figure 2.4: Whiteboards are a great place to centralize ideas when you have more than three people in a room.

Figure 2.5: Always have a good notepad with you to sketch out ideas.

WIREFRAMING

Another phrase you can use to explain your sketches to a customer is the term "wireframing." The term "wireframing" conveys the impression that you are presenting an outline but do not have the substance. It is like looking at the table of contents of a book and using it to judge the value of the content; you get a really good idea but not the details. Still, you do have enough information to make a decision.

TOOL #2: iPAD APPS

How about this for a crazy idea: create an app that lets you sketch apps! Yeah, I know, crazy. But, there are some really cool apps you can use that will let you sketch out designs using your iPad.

OK, so a quick caveat: the iPad apps work on both the iPad Mini and the full iPad, but I strongly recommend you use the full iPad when working on sketches. The iPad Minis screen is just too small.

Tools I use regularly on the iPad are:

- Sketchy
- Penultimate
- Tapose
- Paper

Each of these tools makes it easy to outline the concept of an app and immediately put the design in the hands of the client.

Sketchy

Sketchy is an awesome app for the iPad when it comes to drawing up designs for iPhones, iPads, and websites. The focus of the tool is to give your mock-up a rough, "sketched-out" look but with digital assets. For instance, you will find default layouts for iPhone, iPad, and website apps. In addition, assets such as buttons, image placeholders, media icons, and all of the tools you will need to create a sketch can be dragged onto the main canvas. Objects can be moved around with a touch of the finger and even locked into place. When you have the sketch you like, you can export your design as an image.

Figure 2.6: Knock out ideas using Sketchy.

Penultimate

Penultimate is an illustration tool. The app opens to a blank page that you can edit in much the same way you can a physical notepad. I have found that using a stylus for the iPad gives you great control over your design. The killer feature with Penultimate is that the app is integrated into Evernote's application programming interface (API). This means that you can create the sketch in Penultimate, save to an Evernote book, and then share the book from Evernote with your client. Your client will receive changes as you make them in Penultimate. How about that for communication?

Tapose

Tapose merges the communication of Evernote and the design ability of Penultimate into a single app. The product launched with huge fanfare, but my concern is that the updates have been slow in coming. With that said, the first release is really quite good. You will find cool tools, such as image snipping and rich media, as well as sketching.

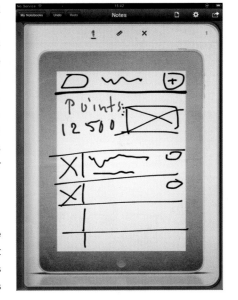

Figure 2.7: Manage your sketches with Penultimate.

Paper

Finally, Paper is an awesome illustrator's app. Like Tapose, it is a tool for creating sketches and keeping the images organized. The difference is that Paper gives you tools that turn the iPad into a canvas to which you can apply digital paint. The result is amazing.

There are many other sketching tools you can leverage for the iPad. These are tools I personally use.

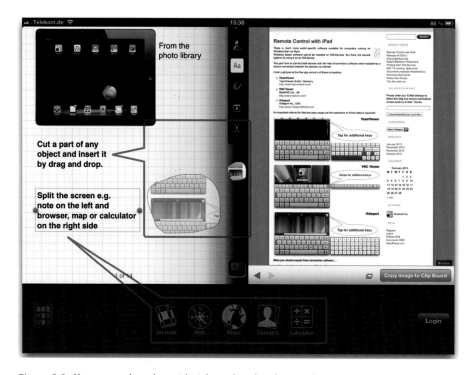

Figure 2.8: Share your thoughts with rich media, sketches, and notes in Tapose.

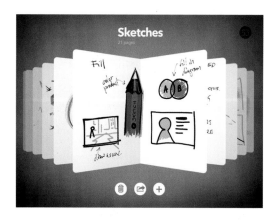

Figure 2.9: Paper gives you lifelike freedom to draw your designs.

TOOL #3: DESKTOP SOFTWARE

There are times I need to go "old school" and crack open my MacBook. The main reason for using my laptop is the greater precision a mouse can give me over my index finger. Also, my MacBook has a 15-inch screen—which is much bigger than my iPad's screen.

The following are tools you can use for creating mobile app sketches.

SwordSoft Layout

SwordSoft Layout is a tool that is similar in concept to Sketchy. You can use it to create sketches that illustrate how a design will come together. What I like about SwordSoft is that it is much more flexible than Sketchy. Also, there are frequent updates that keep adding features and keep the sketching tool relevant on your Mac.

Adobe Illustrator and Photoshop

Adobe Illustrator and Photoshop are two tools you should just have. Period. It is hard to call yourself a designer if you do not have either of Adobe's power tools. Believe it or not,

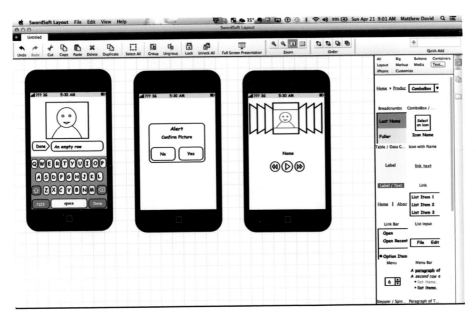

Figure 2.10: Prototype design concepts with SwordSoft Layout.

there is a growing list of awesome prebuilt templates you can download that will instantly give you world-class wireframes built with Photoshop on your laptop.

Omnigraffle

Omnigraffle is a tool that is similar in concept to SwordSoft Layout. The difference is that Omnigraffle really does go the extra mile. You can add flow charts, draft concepts, and build out designs. The difference is price. SwordSoft Layout is only $6.99 and Omnigraffle starts at $29.99 and goes all the way up to $199.99. Bit of a difference.

The bottom line with desktop software is that you do have greater control with a mouse than your finger when it comes to plotting objects onto a sketch. In addition, your sketch can more easily be shared with your client.

TOOL #4: WEBSITES

Every day the Web lets you do more. The fine line between utilizing the functionality of a desktop solution and leveraging the same functionality in a website is becoming blurred. To that point, here are some great sketching tools that you can find on the Internet and that require only a modern browser to use.

Mockingbird—https://gomockingbird.com/

Mockingbird leverages the power of the Web to make sharing concepts very easy with its tools. In addition, you will find the usual suspects for great wireframing tools in Mockingbird. Finally, the price is great. You can get started for free or, if you like what you are using, you can upgrade to the premium package that will set you back only $69 a year. What a deal.

Figure 2.11: There are dozens of great Photoshop templates you can use to design your apps. The template shown is from SpeckyBoy.com.

iPlotz—http://iplotz.com/

A leader in online wireframing is iPlotz. The tool is easy to use and allows you to create a solution from any computer. The iPlotz charge rate is the traditional online "pay as you go" plan that works out to $15 a month.

Figure 2.12: Omnigraffle has many tools that give you control over wireframes, workflow, and design management.

Figure 2.13: Websites such as iPlotz make it very easy to create wireframes without having to buy software.

Cacoo—https://cacoo.com/

Cacoo is like a hybrid workflow and wireframe tool all in one. The price is "free to try" and, if you like the tool, $5 a month. What I like about Cacoo is the focus on building solutions that a whole team can work on, allowing all of the team to share documents and tasks. As the saying goes, it takes a village to raise an app.

Google Drive (Formerly Google Docs)—http://drive.google.com

Arguably one of the most powerful tools you can use to create Web-based designs is Google Drive. Google Drive has been around from many years and is free to use. If you have a Gmail account, then you have a Google Drive account. The tools you can use include workflow diagrams, wireframes, and illustrations. The tool allows you to share your diagrams through e-mail. Google Drive is a great place to start.

At the end of the day, the goal of using these tools, whether it is a pen and pad, iPad app, desktop power tool, or socially aware website, is to give the client a vision for what the app can be.

Applying Human Interface Guidelines to Your Work

How do you, as a member of the human race, use your iPhone? It is not a silly question. The question forces you to think about the user experience, or UX. When I reach for my iPhone, I know I will be using my finger to interact with the screen. I will tap, swipe, pinch, and zoom—all experiences I do in real life. The "experience" is why smartphones are so popular. You do not need to have a Ph.D. to use the device.

The experience is one part of a two-part problem with mobile application design. The second is user interface, or UI. Why is this a problem when all phones are rectangular? The problem is that all phones, indeed all mobile devices, are not created equal. Today there are three leading mobile operating systems: Google's Android, Apple's iOS, and Microsoft's Windows 8.

Each mobile operating system has corresponding hardware. For instance, Apple has only one physical button on the front of the screen, Microsoft leverages tiles in the design, and Google is moving all physical buttons into the virtual world. What complicates issues more is that there is a host of new mobile operating systems that will soon be competing for your customers' eyeballs (BlackBerry 10, Ubuntu, FireFox OS, and Tizen, to name just four).

You need to tailor both UI and UX to the corresponding mobile operating system and to the hardware it is running on.

The design approach is to say, "Well, if it works well on *one* platform then we will keep the design the same for *all* platforms." Let me tell you, as someone who did do that design approach, that it *does not work*.

Let me give an example. A couple of years ago I was developing an education app that performs two functions: it teaches students about ancient Egypt and provides a tool to convert English into hieroglyphics. The first design was for iPhone. Looked great, and today, two years later, I am still getting around 100 downloads per day for the app. Nice. Success.

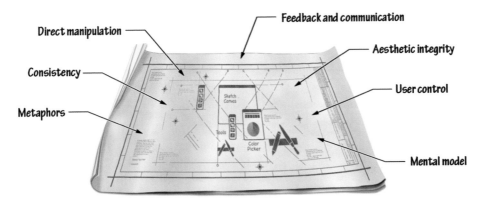

Figure 2.14: Apple's Human Interface guidelines are the gold standard in explaining how we mere mortals interact with complex software.

Then came the iPad and Android versions. I used the same design layout. Total disaster! The reason is that tablets and different operating systems have different user experiences. The result of my bad choice is that for six months we received fewer than 100 downloads. Yes, fewer than I get in one day on the iPhone.

The reason: the App stores require that you present screen shots of what the app will look like. For the iPad, my app looked like I had simply doubled the pixels to fill up the screen, and the Android version was even worse. A typical design technique on Android is to keep your menu buttons at the top of the screen. My menu buttons are along the bottom—yes, it looked like an iOS app on Android, and I was fooling nobody.

So the iPad and Android UI and UX were changed to be appropriate for the corresponding OS and hardware. Guess what happened? Yep, downloads went up almost immediately. Today we are closing in on 125,000 downloads. The Android and iPad downloads now match the downloads for the iPhone. Design for the device UI and UX and don't cheat. People who download your app will know the difference.

THE APPLE WAY

How do you interact with your phone? This is the question Apple engineers asked themselves when creating the iPhone. Indeed, Apple engineers have been posing this question for many years, long before an iPhone was around. "Human interface" is the term Apple uses to describe the experience in designing software. What is great is that Apple has created a set of guidelines you can read and use as you develop your app.

APPLE'S iOS HUMAN INTERFACE GUIDELINES

The iOS Human Interface guidelines are located at https://developer.apple.com/library/ios/documentation/userexperience/conceptual/mobilehig/

The Human Interface guidelines are a set of instructions that take the guesswork out of how you should be designing a solution. Apple pioneered the use of the guidelines with its desktop operating system. The document gives you step-by-step instructions for how to draw out your app design.

Figure 2.15: The iOS team has developed a detailed set of documents you can use to create great iOS apps designed for the iPhone and iPad.

Whether you are designing for a desktop or a mobile operating system, Apple's approach to providing instructions for Human Interface guidelines is to cover the following key principles that provide for a consistent experience for users of Apple products:

- Aesthetic integrity
- Consistency
- Direct manipulation
- Feedback
- Metaphor
- User control

Aesthetic Integrity

Want a beautiful app? Apple doesn't. The goal of aesthetic integrity is not how lovely your app looks but how well the design matches the function. For instance, when you are playing a game such as Temple Run, you want the layout to be fun, but when you are using Evernote to document a meeting, you want the focus to be on your task of documenting the meeting. The Evernote UI should almost appear to fall into the background. The focus is not on lush graphics but on function, not aesthetic integrity.

Consistency

There are a million iOS apps and more than a half billion iPhones. If you want your app to be successful, you need to provide a consistent user experience that can be understood immediately by everyone.

Direct Manipulation

Direct manipulation of the screen through touch, swipe, and gestures is built into iOS. Use these tools where appropriate in your design. The selling point for the iPhone when it was launched was not simply the design. It was the introduction of direct manipulation, demonstrated by Steve Jobs as he swiped the lock button.

Feedback

Feedback is the visual, physical, or audible response to an interaction you perform. For a smartphone, it may be a ringtone when you send a text, open an app, select an icon, or have your phone vibrate when a call comes in.

Figure 2.16: A new era in design, human interface, and computing is launched by Steve Jobs—the start of a multibillion-dollar new market.

When you do something, such as tap a button or complete a task, you are providing feedback to the OS, acknowledging that something has happened. Feedback works both ways. An ingenious feedback tool Apple uses on iOS is the use of the red circles on icons, called badges, that let you know how many e-mails you have not read, how many app updates you have, and so on. Badges drive me nuts (which is their intention) and force me to tap on the icon so I can get rid of the number in the little circle. The feedback I get on the screen is either a change in the number in the badge or its disappearance altogether.

Metaphor

The use of metaphors in an app is to reproduce real-world actions but as virtual experiences. For instance, tapping on the video app opens the app. Inside the video app you can select movies and press virtual play buttons. A scrub bar along the bottom allows you to drag the playback head to the place where you want to start watching your movie. The result is that you make your app easier to learn.

User Control

No one likes it when the computer starts working by itself. For this reason, control should be placed in the hands of the user. Ensure that the user drives actions and tasks, not the iPhone or iPad. If you give the computer control, then we are only a few steps away from HAL singing lullabies.

Apple's Storyboard Technique for App Design

A technique Apple has introduced with the iPhone is known as storyboards. With classic Mac and Windows desktop operating systems, we were all given the option to use multiple apps on the monitor in a "window" metaphor. This gave you the ability to jump from one task to another.

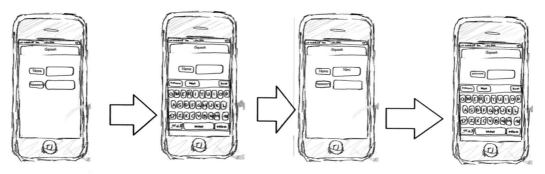

Figure 2.17: You can create a hand-drawn storyboard for the iPhone.

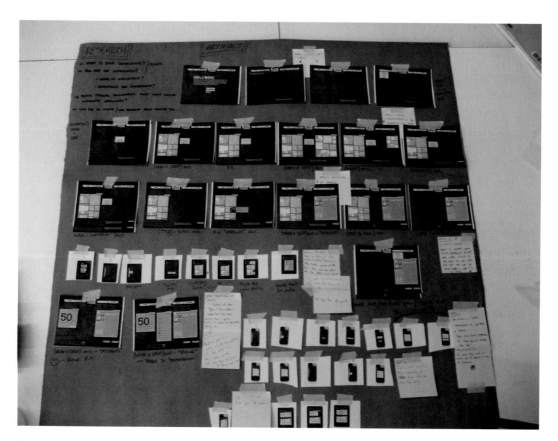

Figure 2.18: Sometimes you need to step back and draw out the concept of an app on the wall with bits of paper so that the whole team can get the idea.

I don't know about you, but it turns out that I am not very good at multitasking. Actually, it makes my work less productive.

Apple believes we are also good at doing only one thing at a time. So let's do that one thing really well.

iOS Storyboards is a simple enough technique. Instead of having a resizable window on the desktop, you have a window that fills the whole screen—no minimize or resize needed. You have just one window with a fixed width and height. The focus is to move from one task to another task, one screen to another. The concept is referred to as a "storyboard."

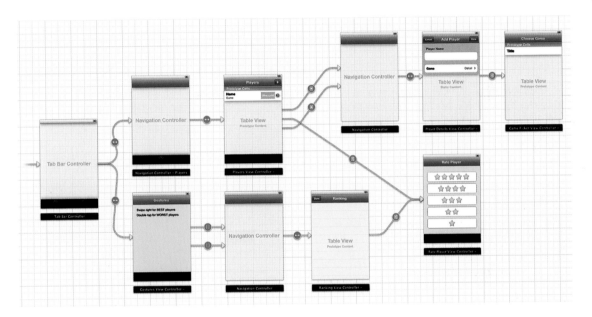

Figure 2.19: From handwritten concept to digital idea, the iOS storyboard technique makes you think about actions rather than windows.

This gives you as a designer a huge benefit. You now have a defined canvas.

Successfully Designing for the iTunes App Store

You do want to pay heed to Apple's advice in its Human Interface guidelines: Apple wrote the book on Apps. Literally.

There are few companies that have put as much thought into what makes a good app as Apple. To that end, it requests you keep the following in mind as you design your app:

- The display is paramount, regardless of size (aesthetic integrity).
- Device orientation can change.
- Apps respond to gestures, not clicks (direct manipulation).
- People interact with one app at a time (storyboard).

- Onscreen user help is minimal.

- Safari on iOS provides the Web interface.

There are now three distinct screen sizes for iOS devices: iPhone (1–4S at 3.5 inches), iPhone 5 (4 inches), and iPad (the iPad Mini is a shrunk-down version of the iPad 2). But Apple has warned that screen sizes can and will change.

You hold your iPhone in your hand and can shake it around. Orientation (Landscape and Portrait) does change. Your app can have a fixed orientation or can reflow the design, depending on orientation. The Stock app is a great example of two different orientations providing different content.

The iPhone and iPad don't use a mouse. They use a finger. What is more, you have more than one finger on each hand. Use them.

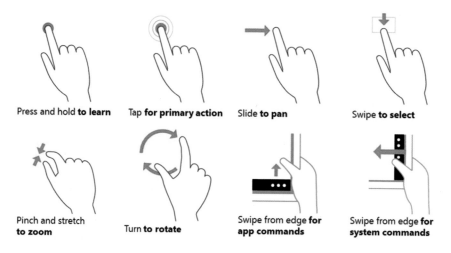

Figure 2.20: Taps and gestures are the primary input mechanism for your apps. Use them!

Apple's use of storyboards for apps does require that one app be visible at a time.

No one likes to have to read a whole book on how to use software. Apps are function driven. If you need to provide training on how to use your app, then you have failed.

Don't convert your website into an app. Make your app an app. The Web can be viewed through Mobile Safari.

The goal when you create your app is to have it published to the iTunes App Store. Apple does heavily regulate the App Store. Abuse of the guidelines is not well tolerated and can prevent your app from being published.

I discuss how to successfully apply the guidelines outlined here in chapter 7.

THE GOOGLE WAY

Unlike iOS, Android is supported on more than 3,000 different devices. Each device has a different specification for hardware, size, screen resolution, and many other characteristics.

When it comes to creative vision, you need to ensure that your solution is built to scale to fit the many different screens.

ANDROID DESIGN

The Google Android design guidelines are located at this website: http://developer.android.com/design.

As with Apple, Google also has its own set of user interface guidelines that Android developers and designers can use. The focus Google places on design falls into the following categories:

- Creative vision
- Design principles
- Style
- Patterns
- Building blocks

Creative Vision

Do you inspire? This is what Google wants you do with Android as part of your creative vision. Today, there are more than 1 million new Android users every single day. You have a massive audience to engage with. Make the engagement inspiring.

Design Principles

Google's design principles fall very closely in line with Apple's and then extend out from Apple's restrictions. In other words, Google encourages you to support human interaction with touch and gestures, but, unlike Apple, Google also encourages almost limitless personalization. Personalization can be slight or extreme. For instance, Facebook's Home replaces the entire default interface to Android with a Facebook-specific solution. Personalization to nth degree! Android is an open-source solution that empowers the developer to extend the operating system in ways that Apple's iOS does not. There are positives and negatives to both approaches.

Style

Android does have consistent styles used throughout the design of the OS. This is common with all operating systems. Common styles include use of Android default themes, grid layout, color, and typography.

Patterns

As with styles, patterns offer you the opportunity to leverage consistent design layout schemes from one Android app to another. Patterns you will want to review and apply to your design include gestures, navigation, action bar, widgets, settings, and notifications.

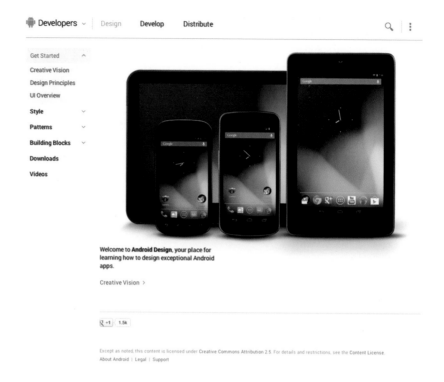

Figure 2.21:
Android is
a flexible
system for
design. Google
strongly
encourages
you to create
amazing apps.

Building Blocks

Finally, the Android OS comes with a set of UI tools you can use as building blocks for your design. Again, the principle is the same as Apple's iOS. Core building blocks are:

- Tabs
- Lists
- Grids
- Scrolling
- Spinners
- Buttons
- Text fields
- Seek bars
- Progress and activity
- Switches
- Dialogs
- Pickers

Google's Android Human Interface guidelines are a tool you can use to create a consistent app experience for your customers. What is different from Apple's system, however, is that you can choose to completely ignore the guidelines and create your own design. It is unlikely that Google will ban your app. Sounds great—creative freedom! But bear this in mind: the guidelines are there to make it really easy for your customers to use your app. Don't make using your app hard. You will lose access to close to a billion customers.

I discuss how to successfully apply the guidelines outlined here in chapter 8.

THE MICROSOFT WAY

Apple and Android dominate the mobile market. Their UI and UX are everywhere. But without doubt my favorite mobile experience is from neither of these companies. It's from Microsoft.

With Windows 8, Microsoft has, for the first time, outdone Apple in user experience. Microsoft has accomplished this stunning feat through a bold directive—to create a smartphone experience that is *not* a traditional Microsoft experience. A new design metaphor, called Tiles, replaces icons; flat images replace detailed icons; and a crisp font is used as part of the design (for those who are curious, the font is Segoe UI).

WINDOWS PHONE

The Windows Phone design guidelines are located at this website: http://dev. windowsphone.com/en-us/design.

Like Apple and Android, Microsoft's design guidelines are split in to principles you can follow. They are:

- Basics
- Principles
- Process
- Library

Basics

Microsoft looked at how packaging and signage are applied in the real world to find inspiration in the creation basic design of Windows 8. This forms the basics of Windows 8. If you want a sneak peak into the mind that made Windows 8 UI, then check out Swiss signage and packaging. The Microsoft team is greatly inspired by the Swiss (maybe they shop at IKEA??).

Principles

Windows 8 leverages grids to drive layout over freeform design as a design principle. You

Figure 2.22: For me, Microsoft's design approach for Windows 8 is ahead of both Apple's and Android's—and Microsoft gives you the secret recipe to replicate its design techniques in your app.

will also see in Windows 8 apps a focus on typography. What font best represents your message? Find it and use it, because it is an important element in Windows 8 apps. Finally, leverage animation built into the OS to reflect interaction with your app.

Process

Microsoft's guidelines define the process you need to go through before you write any code. Define your app's features and focus on the experience; as with Apple, Windows 8 is a storyboard-driven app experience (no windows for you!!). Outline the flow of the storyboard and then apply wireframe designs to each screen. Then iterate through successive designs until you have a solution you are comfortable with.

Library

Finally, Microsoft has an ever-increasing library of documentation you can use.

As with iOS and Android, the Windows 8 experience is unique. The interface is not the same as the competing mobile operating systems. You should not take an app that works great on the iPhone and assume it will port to Windows. The interaction metaphors are just too different.

How to Design with Images

Microsoft has an expression I like when it comes to creating an app: "Speak without words."

Iconography is essential in your mobile design. And iconography is not just images; it is also the use of text as a symbol. Unlike the Web, mobile apps do not restrict you to a fixed set of fonts. A good font can substitute for an image. Think of classic logos such as Nike, IBM, and UPS, which use a font for an image.

CHOOSING THE RIGHT IMAGE FORMAT

There are many image formats you can use. But just use progressive networks graphic (PNG) for your mobile apps. It is supported on all mobile platforms, allows for transparency and photo-rich colors, and is supported on all leading illustration tools.

There, that was easy.

WORKING WITH RESOLUTION

We are in a transition phase in graphics quality in mobile apps. Today, the new devices that are hitting the market come with print-quality screens. Apple calls its system Retina, but we mere mortals know it as PPI, or pixels per inch, where a PPI of 326 or more is the equivalent of a high-quality printed document.

But, there are millions of devices, such as the first-generation iPad Mini, that do not support high-definition resolution images. Even the first devices to hit the market had better than standard Web-image quality. A Web image can have quality as low as 72 PPI. The reason for the low resolution is file size. Denser images are simply much larger and take longer to

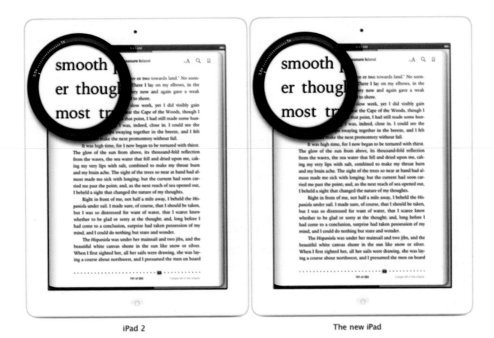

iPad 2   The new iPad

Figure 2.23: The iPad 2 and the new iPads (iPad 3 and 4) have massively different pixel density, as can be seen in the highlighted section.

download over the Web. Fortunately, when you have an image in your app, you do not need to worry about download time. The app is running from the local device.

Here is a list of leading devices and their PPI:

- Acer Iconia Tab—149 PPI
- Amazon Kindle Fire—169 PPI
- Amazon Kindle Fire HD—216 PPI
- Apple iPhone (4, 4S and 5)—326 PPI
- Apple iPad 2—132 PPI
- Apple iPad Mini—163 PPI
- Apple iPad 3, 4—264 PPI
- HTC Evo—216 PPI
- HTC Touch—245 PPI
- HTC One—468 PPI
- Huawei D2—443 PPI
- LG Optimus LTE—329 PPI
- Motorola Droid RAZR—256 PPI
- Motorola Xoom—149 PPI
- Nokia 820—217 PPI

- Nokia 900—217 PPI
- BlackBerry Z10—356 PPI
- Samsung S3—306 PPI
- Samsung S4—441 PPI
- Samsung Galaxy Nexus—316 PPI
- Samsung Galaxy Note II—267 PPI

To this end, you should make your standard images 132 PPI, the screen resolution for the first iPhones and the one used by many entry-level phones.

Applying Responsive Design Layout

The first challenge you have is deciding on image density, the number of pixels per inch. The second challenge you have is the massive array of different screen sizes. Here is the list of mobile devices, with the screen sizes in pixels.

- Acer Iconia Tab—1280 × 800
- Amazon Kindle Fire—1024 × 600
- Amazon Kindle Fire HD—1280 × 800
- Apple iPhone (4 and 4S)—960 × 640
- Apple iPhone 5—1136 × 640
- Apple iPad 2—1024 × 640
- Apple iPad Mini—1024 × 768
- Apple iPad 3, 4—2048 × 1536
- HTC Evo—480 × 800
- HTC Touch—480 × 800
- HTC One—1920–1080
- Huawei D2—1920–1080
- LG Optimus LTE—720 × 1280
- Motorola Droid RAZR—960 × 480
- Motorola Xoom—1280 × 800
- Nokia 820—800 × 480
- Nokia 900—800 × 480
- BlackBerry Z10—1280 × 768
- Samsung S3—720 × 1280
- Samsung S4—1080 × 1920
- Samsung Galaxy Nexus—720 × 1280
- Samsung Galaxy Note II—720 × 1280

You got it. Screen sizes are all over the map. You will notice that the vast majority of the devices in the list run Android; most of the rest run Apple's iOS, and there are a couple of

Windows phones and a BlackBerry device. Creating solutions that can scale to many different screens sizes is critical in your design. Fortunately, there are tools to help.

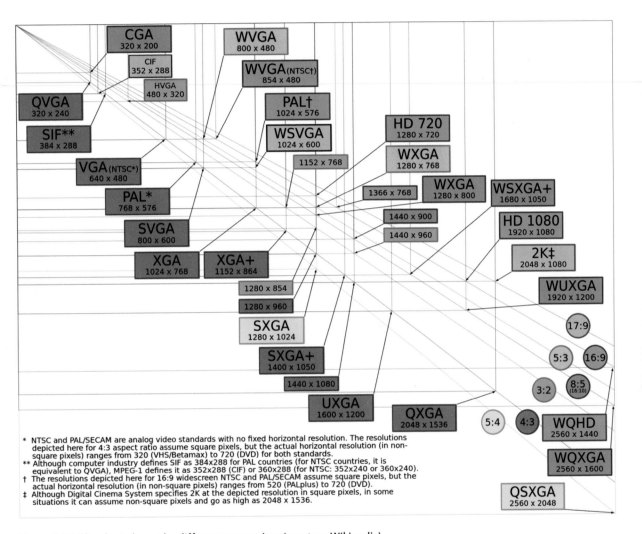

Figure 2.24: The chart shows the different screen sizes (courtesy Wikipedia).

Designing mobile apps is becoming both easier and harder at the same time. It's becoming easier because of the increasing landscape of tools you can use. In the following chapters I present several tools you can use to help with your mobile designs: to create your designs, manage your teams, and build your final digital experience. Tools are here to help.

What is not here? Your focus. Evernote is a great example of a digital solution that runs on all desktop computers, tablets, and phones—an app with great focus, which is to make it easier to record notes in a meeting. That's it. But it does it really well. The other thing

Evernote does really well is to focus how you use Evernote on each device. For instance, the Mac OS X and Windows versions of Evernote are styled differently, because users of those devices work with different OS metaphors; the iPhone and iPad versions of Evernote are different, as one device has a 3.5- or 4-inch screen and the other has a 7.8- or a 9.8-inch screen; finally, the Android version is also different. Use this as an example of how to tailor an app successfully to different screen sizes across the different operating systems.

Design metaphors really matter. The tools are here to help, but it is you who makes the final decisions on how the app is presented. Make sure you focus the design decision around the UI/UX for the device.

From App Planning to Team Creation

The goal of a sketch is to rapidly iterate through designs at a very high level. This is the "blue sky," "out of the box," "anything goes" time in the project. It is the raw sense of idea.

In this chapter we have covered why you want to sketch your app concepts, tools you can use, and methodologies recommended by the leading mobile OS companies. The next step in your app development process is to build your team and start executing the process of creating your solution.

This is where the rubber hits the road.

CHAPTER 3

Defining Your App Creation Team

They say it takes a village to raise a child. The same is true for your apps. You will need a diverse team to build your app the way you want it. In this chapter you will see what types of teams and roles you need to create successful solutions.

How do you define your mobile app team? There are very few apps that are created by one person. Yes, there are success stories such as Doodle Jump (created by a son and a father and still receiving thousands of downloads per day), but for every solo story there are hundreds of stories that talk of large teams working together to build apps.

What you will find as you build out an application is that there are three distinct groups needed to create a successful product:

- The client
- The business engagement team
- The delivery team

Each group is dependent on the success of the others.

The Role of the Client

A client is a purse holder. That's a fact. None of us would be creating apps if it were not for someone paying for the solution. Business stakeholders within your organization can also be seen as stakeholders.

But let's address the real role of the client: visionary.

The client is the person who understands the potential of what can be done. She may not know how to complete her vision, what the cost will be, or how to get there, but she does have a blinding vision.

The business world is littered with leaders with strong visions: Steve Jobs, Bill Gates, Mark Zuckerberg, and Jeff Bezos, to name four. They saw a place that was not currently being occupied and understood that a vital new product was needed to fill the void: a mobile device, personal computing, a facilitator of immediate personal connections, and a radicalization of retail shopping.

These four leaders are known the world around, but we all know visionaries who see the world differently. Visionaries are seeing mobility as the next digital frontier. However, there is a fly in the ointment with the release of mobility as the next "big thing." It is a technology change that is coming with two kissing cousins: cloud and social media.

The cloud is a euphemism for data storage, solutions, and flexible systems that are run by a third party on the Internet (or in the Internet cloud). There are several leaders in the cloud space: Amazon, Microsoft, Google, and SalesForce. Without getting too into the details, cloud solutions empower entrepreneurs to walk with the big boys. You do not need to build an expensive data center when you leverage cloud-based solutions. For pennies on the dollar, you can leverage cloud product to have the same technology infrastructure as a Fortune 500 company; speed the implementation of complex technologies (for instance, you can be up and running with Office365 for e-mail, calendar, SharePoint, and Lync for communication in less than an hour). Cloud systems give you the option to reduce the barriers of cost and startup time that otherwise plague startups.

Social media is still changing our lives. We have Facebook, Twitter, Google+, and more choices every day. What social media are doing is breaking down the walls around us. The line between work and play is getting thinner every day. For instance, part of my job is to be active on Twitter, discussing mobile design and HTML5. But my audience is scattered all over the world. It is not unusual for me to be tweeting with someone at 10 in the evening. I'm still working (technically) but having fun, too.

Your clients see mobility, cloud, and social media as tools they can use. An app is the forward-facing part of your mobile engagement with your client. Use your app designs to illustrate cross-platform design (for broad device support), cloud storage (for cost effectiveness), and social media (for reach).

CHOOSING AMONG OFF-SHORE, ON-SITE, AND NEAR-SHORE

The client holds the keys to the kingdom when it comes to managing costs in running a project. The biggest cost will come in the development of the solution you are building. There are three traditional methods for developing solutions. They are:

- Off-shore

- On-site

- Near-shore

The off-shore model for solutions was pioneered in the 1990s with the massive growth of companies in India that were able to provide highly skilled resources to work on technical solutions for companies in Europe and North America.

The single largest benefit of off-shore is price. Even today, the price to hire an off-shore team is typically a third of what you would have to pay a skilled resource in the United States.

Of course, for every benefit there are detriments. Stories you will hear talk about significant time zone differences, cultural differences, and communication challenges. The list goes on and on. Frankly, I don't buy it. These just seem like excuses to me. When you are working with any team, you have to manage the team. The same is true with off-shore. Do not expect to throw work at a group and expect the group to deliver a perfect solution. You will need to manage the off-shore group as you would a local team.

The traditional method for delivering solutions has been by hiring or contracting with a person to work on-site for the client. That means the resource you contract is sitting down at a cube and working with company employees. It is hard to distinguish who is the contractor and who is the employee.

Again, there are benefits and detriments. The benefit is that you get to see the work being done. You can keep your eye on that very expensive resource you have contracted. The detriment is that it is tempting to have the resource work on other projects. The person is just there, after all.

The final development model you can leverage is one that has been used in the marketing world for many decades: near-shore. The concept for marketing agencies is this: we have the talent (or access to the talent) that a chief marketing officer requires. The CMO can buy the agency's talent and have them complete work on a project-by-project basis. It is the role of the agency to manage the resources it has and to augment the team as it sees fit. The client does not see any of the activities that go on behind the scene: all it sees is the final product.

I believe the near-shore model will become the default model moving forward as the CIO supplements its core team with highly skilled delivery groups that manage resources at a different location.

THE COMMON ELEMENT: PEOPLE

If there are three modalities for managing teams, the common element for all three is this: people.

Teams that are motivated are successful. Positive thought engages positive actions. You are the client, and you are the leader. You are paying the bills, making the final design choices, and marketing the final product. So see yourself as the leader of the group. Be the person of change who motivates the team.

Complex and long projects lead to resource fatigue. The client's role is to keep the team motivated to deliver a solution to change the world.

The Business Engagement Team

Before a project can be worked on, a team has to convince a prospective client that it should part with its money and spend it on your delivery team. This task is the domain of the business engagement team. And business engagement is not just sales.

Believe it or not, we are not in the *Mad Men* world anymore. Today, a sale is not completed by fraternity-like drinking contests, trips to a "Gentleman's Club," or intimidation.

Effective sales teams are not about selling you a used car and hoping you don't read the small print.

Today, the team that engages with the client needs to be effective at the following:

- Understanding the client's challenges (both external and internal)
- Clearly defining the method of delivery for the solution
- Managing the client's expectations throughout the app creation
- Being part of the client's team to win success

The bottom line is that the business engagement team is the communications link between the client and the delivery team. Its goal is to be the pivot for a teeter-totter between the client and the delivery team.

The Delivery Team

The good news is here: you have won the contract to build an app. Now the rubber hits the road. The delivery team must make good on the promise of delivering a solution that will delight the customer. There are essentially two main schools of thought when it comes to managing project delivery: Waterfall and Agile.

The Waterfall approach requires a project manager who coordinates a large, disparate team over a period of several months to complete the work. The project is often broken into major milestones that can include the following:

- Project acceptance
- Project definition
- Project kick-off
- Requirements gathering
- Phases 1–4 for product milestones
- Quality control
- Release
- Retrospective

These milestones each take anywhere from two to eight weeks to complete. As you can see from the list, you thus have a project that takes the better part of a year to complete.

Agile, a methodology that has its roots in Lean Manufacture, takes a very different approach. Instead of separating skill competencies by milestones, you put all the people with the correct skills together and have them work through two- to four-week "sprints" to complete a set of tasks that will produce a complete product. Yes, you heard me, at the end of a sprint you have a product that works or clearly defined groundwork for future sprints. The product may not have a lot of features, but it is functional.

The idea with the Agile approach is that you keep the teams together in small groups that keep iterating through tasks until you are ready to launch the product.

Both approaches have had success, and both have had failure. At the end of the day, you must focus on the success of the project first and foremost. Use project management tools to get the job done, but make sure you do get the job done and do not get lost in the detritus of project management and execution.

WHY WATERFALL WORKS FOR THE PURSE HOLDER BUT NOT FOR THE TEAM

As the client investing in a project, all you want to see is the final product: your app. You are willing to pay $100,000 for the app. You want to know how long it will take to deliver the app, and you want to know when you can start making money from it.

The Waterfall approach is great. You spend time at the beginning of your project getting all of your requirements to the team. This allows you to leave and wait for the final product.

But this approach does not work for the delivery team. Additional questions will come up: what color do you want this button; how should this feature react when in XYZ scenario; and why do you want that!?

The result is that as many as 68 percent of waterfall projects will fail, according to a study, "The Impact of Business Requirements on the Success of Technology Projects," by IAG Consulting. You can read the report, but it comes down to one critical issue: communication.

Continuous communication with a clearly defined cadence is critical in the success of any project. This is why you are seeing companies adopting a new delivery technique: Agile.

CHOOSING AGILE TO KEEP ON TARGET AND DELIVER ON BUDGET

Over the past decade, a new methodology for delivering solutions, called Agile, has emerged. Agile breaks with many of the tenets of Waterfall. The key elements to Agile are:

- Small, cross-functional, empowered groups made up of developers, QA, business analysts, and architects
- Short two- to four-week "sprints" with a clearly defined and workable solution at the end of each sprint
- Client as part of the delivery team
- Use of whiteboards to post tasks and keep unnecessary documentation to a minimum to adapt to change as it happens

Agile flies in the face of Waterfall. But it really works, and it gets the job done. Agile has two great advantages over Waterfall: the client is always involved (to the point where many clients join the daily "standup," a 15-minute run-through each morning of the day's activities, highlighting roadblocks), and the client gets to use a workable solution at the end of each sprint.

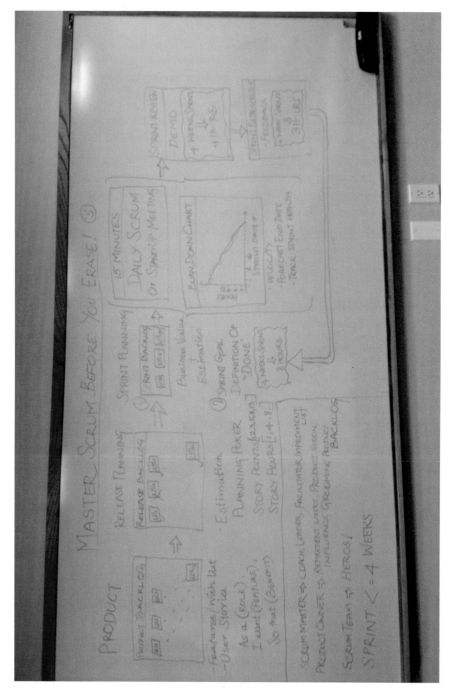

Figure 3.1: Scrum is a variation of Agile that uses two-week sprints to complete tasks.

THE TWO-PIZZA RULE

A core theme to any Agile process you adopt is small, empowered teams. The focus is that just enough people have been brought from various competencies to get the job done. The team is nimble and scrappy. But how do you know you have too many people in your Agile team? Use the Two-Pizza Rule. If you need more than two pizzas to feed the team, then you have too many people.

THE AGILE MANIFESTO AND 12 PRINCIPLES

To learn more about Agile, visit the website Agilemanifesto.org.

Agile was defined during a meeting on February 17, 2001, at Snowbird, Utah. The 17 participants were there to discuss lightweight development methodologies, and from this discussion was produced the manifesto that is the core for Agile as we know it today:

We are uncovering better ways of developing software by doing it and helping others do it. Through this work we have come to value:

Individuals and interactions over processes and tools

Working software over comprehensive documentation

Customer collaboration over contract negotiation

Responding to change over following a plan

That is, while there is value in the items on the right, we value the items on the left more.

And so, a new way to get what we need done was agreed upon.

ITERATIVE RELEASES FOR VISUAL SIGNS OF PROGRESS

Early in my career, a good friend of mine told me something: it doesn't matter how much you do if you are not providing visual signs of progress.

Took me a while to figure out what he meant by his Yoda-like advice. Turns out, Agile is really all about visual signs of progress. Your delivery should focus on providing constant feedback to the client about the work you are doing and how the client can contribute. Do not work in a vacuum. This only creates anxiety.

More recently I was given additional advice: no one likes surprises. This is why we always hunt for our presents at Christmas.

Keep the lines of communication open. Tell your client what you are doing, and ensure that you reach out and receive feedback. You will provide clear and consistent signs of progress, and your client will not get a surprise when presented with the final product.

NEMAWASHI—TO WORK AROUND THE ROOTS

No project is complete without conflict. It is human nature. Put two people in a room and you will get two viewpoints. Put three people and you will get five (not sure how the math works here, but try it, and you will get five viewpoints). So how do you get everyone to agree to one viewpoint? The Japanese have a concept, *nemawashi*, that I have used to get everyone in the room to agree to one viewpoint. *Nemawashi* means to work around the roots. For your project, this means working 1:1 with each team member to present the conflict, assess different viewpoints, and present your own viewpoint. The idea is simple: meeting with each person before you get into the conference room means you know all the viewpoints that can seed the single viewpoint you want to succeed. From here you can grow a tree. You are, to all intents, working around the roots of the problem.

The Roles

We all play different roles in the delivery of a project. Whether you choose Agile or Waterfall, you will need experts who can take your ideas and bring them to life. You need your delivery team. Your work will need a team of skilled resources for it to be successful.

BUSINESS LEADERS

Earlier in the chapter I talked about the importance of having the client involved in all phases of the project. This includes the delivery team. Either the client or the client's authorized representative needs to be part of the delivery team to ensure that the product continues to meet the initial vision and can go to market with a big impact.

DESIGNERS

The previous chapter covered the exciting world of wireframing and app concepts. A designer is involved heavily in creating these concepts. But the role does not stop there. Many designers are now kept closely involved throughout the development of the whole project. The reason for this is the high UI/UX complexities in mobility created by Apple, Google, and Microsoft. Indeed, many fine-art schools are now teaching programming; among these is the Academy of Art Institute (where I teach a master's class in mobile app design—shameless plug).

ARCHITECTS

In my opinion, the most important person on the delivery team is the architect. It is the architect's role to define how the architecture will be created to scale to massive demand (it is not uncommon for mobile apps to see thousands of downloads in days) and to ensure that you are developing the solutions for the right platforms.

CLIENT DEVELOPMENT

Client development does not refer to the person with the money. We are now going hard-core programming and talking about the mobile client, the operating system, and the hard-ware. There are four major client environments you need to be sure to cover. They are:

- iOS
- Android
- Windows
- HTML

The next section of the book is going to deep dive into the tools you can use to develop solutions for these platforms. Each client is different and comes with its own challenges. Some people are terrified of challenges. Mobile developers are a different breed. They say: bring it on!

Quality Management

Customers catch poor quality in a product very quickly. Walk into a store and choose a banana. Do you pick the perfectly yellowed fruit or the fruit with the black blotches? One banana will last and one needs to be made into banana bread today (tomorrow will be too late).

It is the role of quality management to ensure that you have the best possible product delivered to your customer. Earlier I discussed Waterfall and Agile as two ways in which you can deliver solutions. There are variants of both methodologies. What is critical, however, is clear communication in the team. Here are some techniques I use that have been effective:

- Sprint 0—project definition: meet for a two-week period at the beginning of the project to define the larger scope of the final product that will be released. You can complete only so many features in each sprint, and you will need several sprints before you have a deliverable solution.

- Daily standups—take 15 minutes at the beginning of each day for a standup meeting (standing keeps everyone itchy to get done) to identify what has been done in the past 24 hours, what will be completed in the next 24 hours and, most important, what roadblocks will prevent you from getting your work done.

- Daily summaries, in bullet points, of the work done—three to five bullet points should be more than enough for each person.

- Weekly summaries for the client with a traffic signal light (red, amber, green) against each major feature for the work period to indicate concern, potential problem, or on track.

- Retrospectives at the end of each phase to review what has been learned and what can be done better.

Leveraging these tools can help keep the lines of communication open in the team.

KAIZEN—CHANGE FOR THE BETTER

The Japanese manufacturing industry uses a term, *kaizen*, in daily life. Simply put, *kaizen* means continuous improvement. In essence, it means, What can we do today that is better than what was done yesterday?

Leverage the philosophy of *kaizen* in your team. Make today better than yesterday through continuous change.

Changing the World

You are now in an amazing place. You have a client, a business engagement team, and a delivery team. You understand that a mobile experience is different from a Web or desktop solution. You are moving into new territory.

The mobile evolution is currently in its infancy. The work you now do will be groundbreaking. You can change the world. Infuse your team with this single belief: you are not creating an app; you are changing someone's life.

In the next section of the book, we dig into the tools you can use to create your world-changing solutions.

SECTION

2

Building Apps

Creating Apps with Adobe PhoneGap

Adobe's PhoneGap is now the de facto hybrid technology platform to kick-start your native app creation process. The secret to PhoneGap is HTML5. In this chapter you will be introduced to cross-platform app development with PhoneGap and learn how you can leverage HTML5 to create solutions.

Building Apps with PhoneGap

Do you want to build an app? Of course you do. Everyone does. Apps are the next cool technology, and we all want to play with the cool technology.

There is a roadblock when it comes to creating mobile apps: too many new technologies. For instance, iPhone and iPad apps are developed with Objective-C, Android apps are developed with Java and C++, Windows 8 apps are developed using C#. The list of technologies you need to learn today as a developer is staggering.

So how do you make the process of creating apps much easier? You look to forward-thinking technologies and bet hard they will work.

Figure 4.1: Adobe's PhoneGap is a cross-platform mobile OS framework you can use to build native apps.

PhoneGap is a framework that is looking to reduce the level of complexity we currently see in mobile app development through leveraging HTML5 as a cross-platform development language. The key to PhoneGap's success is that you can build your entire solution using Web technologies you are already familiar with—HTML, CSS3, JavaScript, SVG, and Canvas—without having to learn complex platform-specific technologies.

Figure 4.2: Use HTML5 to power your apps in PhoneGap.

PhoneGap is an open-source project (now called Cordova, but the term PhoneGap and Cordova are almost interchangeable). You can download the entire code from a GIT repository and build the solution yourself. Adobe is the company that manages the project.

The goal of PhoneGap is to allow you to turn HTML5 apps or micro-sites into native applications. This is done two ways:

- Compiling your PhoneGap app on your development machine
- Leveraging Adobe's cloud-based PhoneGap build environment

Both solutions give you the opportunity to create apps for ten different platforms. With one set of HTML5 code you can create native apps for the following:

1. Android
2. iOS
3. Windows Phone 7
4. Windows Phone 8
5. Windows 8
6. BlackBerry
7. Tizen
8. Bada
9. Symbian
10. WebOS

What is exciting, as you develop solution with PhoneGap, is that you can leverage PhoneGap's ability to connect to native APIs from the device operating system and, through JavaScript, connect to the Camera, Contacts, Events, GeoLocation, Notification, and many of the features typically reserved for native development.

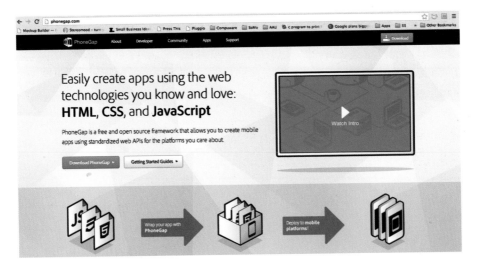

Figure 4.3: PhoneGap's website.

Close to one in four hybrid apps is now developed with PhoneGap. It is easy, supports many platforms, and appears to be a silver bullet for rapid cross-platform development. The center of PhoneGap is HTML5.

ONE TECHNOLOGY TO RULE THEM ALL—HTML5

HTML5 is an awesome technology. I know; I've written three books on it. I think it rocks!

But here is the secret: it is not that new. At least, it is not as new as the press would have you think. HTML5 is a clever brand tying together a collection of technologies and putting forward a solution stack that can create apps that behave like native desktop solutions.

The core to HTML5 is the following technologies:

- HTML5 Elements
- CSS3
- Video and Audio Elements
- SVG and Canvas
- JavaScript

The core elements of HTML are the tags you see in your page. They include but are not limited to the following elements:

- BODY
- HEAD
- SCRIPT
- HTML
- P
- IMG

You will see the tags displayed on the page in the following format:

```
<HTML>
<HEAD>
<TITLE>This is a Web Page</TITLE>
</HEAD>
<BODY>
<P>Hello, World</P>
<IMG SRC = "sample.jpg">
</BODY>
</HTML>
```

The framework for HTML elements (originally created by Tim Berners-Lee in 1989) was published in 1993 and took the world by storm.

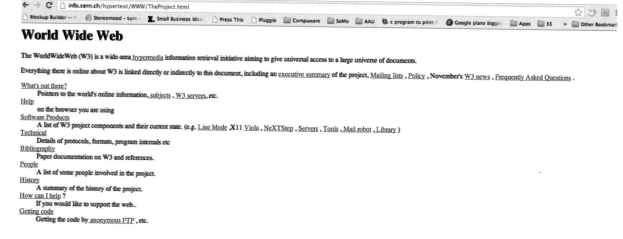

Figure 4.4: A copy of the very first website. We have come a long way in the past 20 years!

But, the original spec was not perfect. Over the next four years, three more iterations of HTML came out, adding in more complexity (notably the inclusion of JavaScript to add client-side interaction), but development of HTML appeared to slow down by the end of the decade. In 2004, a group known as Web Hypertext Application Technology Working Group (WHATWG) began work to bring consistent Web technologies to all. The pioneer of WHATWG is Ian Hickson (who now works at Google). It is through his work and the consolidated work of the team at WHATWG that lesser-known technologies, such as CANVAS, SVG, and CSS3, were brought together and presented as HTML5.

Figure 4.5: The WHATWG started evolving HTML in 2004.

The adoption for HTML5 has been nothing short of amazing. Today, all modern Web browsers support most of the features outlined in the HTML5 specification. This includes Google's Chrome, Apple's Safari, Mozilla's FireFox, and Microsoft's Internet Explorer.

What is interesting about HTML5 is that it is not a "cast-in-stone" specification. The modular design of the specification is intended to allow emerging technologies to be integrated quickly. For instance, when HTML5 was first published, we did not have smartphones or tablets as we know them today.

To this point, some of the first Web browsers to have the best support for HTML5 are mobile devices such as iPhones, iPads, Android phones, WebOS, and BlackBerry OS. Each of these platforms has technology that allows you to embed a Web browser window into a native app and present HTML pages through it. PhoneGap is leveraging the cross-platform support for HTML5 as a cornerstone in building out sophisticated apps. Where a mobile OS does not support a feature in HTML5, such as a Camera API, then PhoneGap will extend the browser with JavaScript into the native operating system. Clever stuff.

Geolocation	X	3.0	2.0	6.0
DOM Storage (noSQL)	7.0	3.0	2.0	6.0
Web workers	X	X	2.0	6.0
Web Socket	X	4.2b	X	X
Canvas	X	1.0	1.0	6.0
CSS 3.0	X	3.0	2.0	X
App Cache	X	2.0	2.0	6.0
Device (camera, etc)	X	X	3.0	X
Touch Events	X	2.0	1.0	X
Audio/Video	X	1.0	2.0	X

Figure 4.6: HTML5 offers support for mobile browsers.

Today, HTML5 is not limited to just Web browsers on your desktop or mobile device. New implementations of HTML5, through projects such as FireFox OS, Chrome OS, WebOS, and Tizen, are changing HTML5 into the de facto native programming language for the next generation of operating systems.

And, yes, PhoneGap is already working on these forward-thinking HTML5 operating systems.

THE PROBLEM PHONEGAP IS SOLVING—CROSS-PLATFORM APP PUBLICATION

Adobe is set with a big challenge: how to boil the ocean. What do I mean by that? The phrase "boil the ocean" refers to massively complex projects that tackle massively complex problems.

Normally, when you hear someone say "that project is boiling the ocean," it means that the project is being overengineered.

Not the case with PhoneGap.

The challenge Adobe is addressing with PhoneGap is massive: to create a single authoring language common to close to a dozen disparate operating systems. Sun Computer tried it with Java, and it did not work. So why is Adobe trying?

It is the support for HTML5 on all the mobile devices that is making Adobe's job much easier. Apple, Microsoft, Google, Mozilla, Opera, Oracle, IBM, and many companies are all working to make HTML5 the cross-platform technology of the future.

Phew! Seems like job done, doesn't it? Well, not exactly. While HTML5 is great, it does not address specific features found on native operating systems. A native operating system is generally controlled by a single, private entity and can effectively do what it likes. This is clearly seen with Apple and its iOS platform, used to run iPhones, iPads, iPod Touches, and Apple TV.

To begin with, iOS has three main different user experiences: iPhone/iPod, iPad, and Apple TV. The input mechanism for each is different: single finger, two hands, or a remote control. HTML5 does not fully support gesture controls (e.g., taps, two-finger swipes) and does not support core APIs built into iOS. For instance, iOS has the following APIs available only for native application developers:

- MapKit—for adding maps to your projects
- GameKit—adding interactive game features such as accomplishments
- iAd—for including advertising in your apps
- Passbook—Apple's digital wallet solution
- Notifications—those annoying alerts you get on your iPhone
- In-App Purchases—ways to extend your app with additional, paid-for features
- Camera—Lets you control your camera and Camera Roll

This is just a small collection of features that HTML5 does *not* support. This is where PhoneGap gets really cool. Through PhoneGap you can build out the majority of your project in HTML5 and re-use it across multiple platforms. But, if you need to add a specific OS feature, you can branch your main project and include, through either JavaScript or native code, the feature you want. For instance, the Accelerometer is not an HTML5-supported feature. But, through PhoneGap and its use of JavaScript to connect to native features in a mobile operating system, you can add the following HTML5 app and use it in any language supported by PhoneGap:

```html
<!DOCTYPE html>
<html>
<head>
<title>Acceleration Example</title>
<script type = "text/javascript" charset = "utf-8" src =
"cordova-x.x.x.js"></script>
<script type = "text/javascript" charset = "utf-8">
document.addEventListener("deviceready", onDeviceReady,
false);
function onDeviceReady() {
navigator.accelerometer.getCurrentAcceleration(onSuccess,
onError);
}
function onSuccess(acceleration) {
alert('Acceleration X: ' + acceleration.x + '\n' +
'Acceleration Y: ' + acceleration.y + '\n' +
'Acceleration Z: ' + acceleration.z + '\n' +
'Timestamp: ' + acceleration.timestamp + '\n');
}
function onError() {
alert('onError!');
}
</script>
</head>
<body>
<h1>Example</h1>
<p>getCurrentAcceleration</p>
</body>
</html>
```

A second example is leveraging the Notification API for a mobile OS. Again, the concept of notifications is not supported by HTML5, but the following example shows you how to place notifications into your projects:

```html
<!DOCTYPE html>
<html>
<head>
<title>Notification Example</title>
<script type = "text/javascript" charset = "utf-8" src =
"cordova-x.x.x.js"></script>
<script type = "text/javascript" charset = "utf-8">
document.addEventListener("deviceready", onDeviceReady,
false);
function onDeviceReady() {
}
function alertDismissed() {
}
function showAlert() {
navigator.notification.alert(
'You are the winner!', // message
alertDismissed, // callback
'Game Over', // title
'Done' // buttonName
);
}
</script>
</head>
<body>
<p><a href = "#" onclick = "showAlert(); return
false;">Show Alert</a></p>
</body>
</html>
```

The value you can see here is that PhoneGap is augmenting the cross-platform support of HTML5 with hooks into native features. This is not an easy task, and Adobe is boiling the ocean to bring an excellent tool for you to use on any platform your client asks you to support.

WORKING WITH CORDOVA

You will hear PhoneGap referred by another name: Cordova. Cordova is the name Adobe gave PhoneGap when it acquired the technology, in 2011. Adobe wanted to keep Cordova as an open-source project, but, due to legal reasons, it was required to change the name. So the name Cordova was born.

Figure 4.7: The Apache Cordova project is the open-source name for PhoneGap.

No, I have no idea what Cordova means. Sounds like a great coffee.

If, however, the idea of working with a native authoring environment gives you wobbly knees, then Adobe has a cloud tool, PhoneGapBuild, to which you upload your apps. This is the location of the website: Build.PhoneGap.com.

If, however, you are a hardcore developer and want to dig through "pieces" that make up the source code used to create PhoneGap, then you need to check out the following site: https://github.com/phonegap.

But, to be clear, the GitHub repository of PhoneGap is not for the weak-kneed.

So, whether you call it Cordova or PhoneGap, you still have a collection of very cool tools at your disposal.

DEVELOPING WITH HTML5

To be fair, you really do not want to use HTML by itself to create your apps. It will be a lot of work. Fortunately, there are large collections of frameworks you can leverage to create native-looking apps. The list includes the following frameworks built with JavaScript, HTML, and CSS to rapidly develop solutions:

- jQuery Mobile (www.jquerymobile.com)
- Sencha Touch (www.sencha.com)
- Twitter Bootstrap (http://getbootstrap.com/)
- Kube (http://imperavi.com/kube/)

Using a framework will speed up the development cycle you will go through to deliver rich and compelling apps. I am going to zero in on my favorite framework, jQuery Mobile, as an example.

WHO IS USING FRAMEWORKS?

Frameworks such as jQuery Mobile get you up to speed very quickly. Need convincing that frameworks are a smart move? Then look at the following companies that are already using jQuery Mobile:

- Disney World
- SlideShare
- Box.Net
- IKEA
- Stanford University
- Khan Academy
- Rugby World Cup
- OpenTable

There are many more examples, but these are some brands you are likely familiar with. Their support of jQuery Mobile should help you sleep at night knowing you have selected a great product.

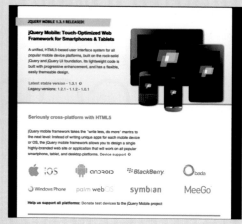

Figure 4.8: jQuery Mobile runs on all leading mobile and desktop browsers.

Let's take a simple Web page.

```
<html>
<head>
<title>
Sample Page
</title>
</head>
<body>
<H1>A Link</H1>
<a href = "#">Click me</a>
<H1>Lists</H1>
<ul>
<li>Red</li>
<li>Green</li>
<li>Blue</li>
<li>Yellow</li>
<li>Orange</li>
```

```
</ul>
<H1>Using a Form</H1>
<form>
<label>Text Input:</label>
<input type = "text" id = "textinput-s" placeholder = "Text
input" value = "">
<label>Select:</label>
<select>
<option value = "small">One</option>
<option value = "medium">Two</option>
<option value = "large">Three</option>
</select>
</form>
</body>
</html>
```

View the Web page and it will not look very exciting in any Web browser.

A Link

Click me

Lists

- Red
- Green
- Blue
- Yellow
- Orange

Using a Form

Text Input: Text input   Select: One

Figure 4.9: A standard HTML5 page with no formatting.

Now, let's add a little jQuery Mobile magic. In the HEAD section of the HTML page, add the following:

```
<link rel = "stylesheet" href = "http://code.jquery.com/
mobile/1.3.1/jquery.mobile-1.3.1.min.css"/>
<script src = "http://code.jquery.com/jquery-1.9.1.min.js"></script>
<script src = "http://code.jquery.com/mobile/1.3.1/jquery.mobile-
1.3.1.min.js"></script>
```

These files, one CSS style sheet and two JavaScript documents, link to the public jQuery Mobile and jQuery files.

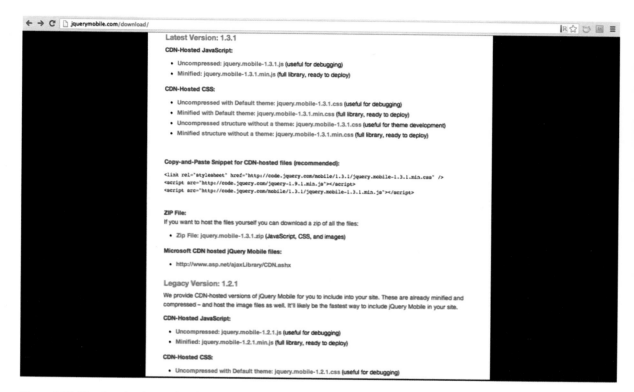

Figure 4.10: The download page for jQuery Mobile.

There is a concept known as "pages" in jQuery Mobile that lets you create multipage solutions from one document. Let's add a "page" to the HTML just shown.

Immediately after the BODY element add the following:

```
<div data-role = "page">
<div data-role = "header">
<h1>Single page</h1>
</div><!—/header—>
<div data-role = "content">
```

Locate the closing BODY element at the bottom of the page and add the following:

```
</div><!—/content—>
<div data-role = "footer">
<h4>Footer content</h4>
</div><!—/footer—>
</div><!—/page—>
```

At this point, you have created a basic page that is optimized for a mobile device. You can test the page in any Web browser to see what it looks like. You will see that the page now has a theme, footer, and header as well as a mobile-like design.

OK, that's cool, but let's dig a little deeper. Take a look at the Web link added to the page:

```
<a href = "#">Click me</a>
```

In the A element add a new jQuery Mobile–specific attribute called a data-role and a stylistic attribute called a data-icon as shown here:

```
<a href = "#" data-role = "button" data-icon =
"star">Click me</a>
```

Save your file and preview it in your browser. WOW! The link now looks like a button you can press with a finger. It is still just a URL link, but you have now optimized it for a mobile device.

Lists can also be easily changed. Here is the HTML code for a simple list:

```
<ul>
<li>Red</li>
<li>Green</li>
<li>Blue</li>
<li>Yellow</li>
<li>Orange</li>
</ul>
```

Figure 4.11:
jQuery Mobile
transforms the
Web page into
a solution that
is optimized for
a mobile device
screen.

Radically change the appearance and features of the list by adding the following attributes to the opening UL element:

```
<ul data-role = "listview" data-inset
= "true" data-filter = "true">
```

View the page in your browser. Now you can easily scroll up and down the list by using your finger and filter it with an input field. All you had to do is to add the following attributes: data-role = "listview" data-inset = "true" data-filter = "true."

Finally, let's look at a FORM.

```
<form>
<label>Text Input:</label>
<input type = "text" id = "textinput-s" placeholder = "Text
input" value = "">
<label>Select:</label>
<select>
<option value = "small">One</option>
<option value = "medium">Two</option>
<option value = "large">Three</option>
</select>
</form>
Compare with the few additional attributes added below:
<form>
<label for = "textinput-s">Text Input:</label>
<input type = "text" name = "textinput-s" id =
"textinput-s" placeholder = "Text input" value = "" data-
clear-btn = "true">
<label for = "select-native-s">Select:</label>
<select name = "select-native-s" id = "select-native-s">
<option value = "small">One</option>
<option value = "medium">Two</option>
<option value = "large">Three</option>
</select>
</form>
```

It is the same form, now optimized for a mobile device. You can even take forms one step further with jQuery Mobile and add widgets such as sliders. The following will add a slider that you can drag back and forth with your finger:

```
<label for = "slider-s">Input slider:</label>
<input type = "range" name = "slider-s" id =
"slider-s" value = "25" min = "0" max = "100"
data-highlight = "true">
```

Figure 4.12: Advanced UI features such as Themese and Sliders are supported by jQuery Mobile.

The end result is that you can create complex-looking solutions with frameworks such as jQuery Mobile.

The goal in using PhoneGap is to create cross-platform solutions that use HTML5 technologies as a common language. But, PhoneGap does give you a "get out of jail free card" and offers the opportunity to dig deeply into core APIs for a mobile OS.

So, let's dig into iOS development within PhoneGap and see how you can extend the functionality beyond basic HTML5 support.

Building for iOS with Cordova

Apple did a clever thing when it released iOS. The core of iOS is branches of the same core found in Mac OS X. Why create a totally new OS when you already have a great one?

The advantage in building out on an existing operating system is that you can leverage existing APIs and development environments. When Apple announced the App Store in 2008, it also announced that any developer who was already familiar with the authoring environment for Mac OS X would be able to get up and running with iOS very quickly.

Apple's authoring environment, called Xcode, has been used to create solutions for the Mac since its release in 2003. Through Xcode you can visually design your apps, introduce functionality through programming, test and debug your solution, and submit it to the iTunes App Store. Indeed, Xcode is a powerhouse of tools.

Figure 4.13: Xcode is the design environment you will use to build native PhoneGap iOS solutions.

Central to the development you will do in Xcode is the programming language Objective-C. Objective-C has its roots in C++ and SmallTalk. Often referred to as Obj-C, the language is both mature (the first published version of Objective-C was in 1986) and stable. The language is a true object-oriented-programming (OOP) language that allows teams to develop highly scalable solutions. We will dig more deeply into Xcode and Objective-C in chapter 7.

There is a big challenge in working with Objective-C: not many people know it, and it is very complex to learn.

Figure 4.14: Developing with Objective-C in Xcode for the iPhone.

For this main reason, many App developers, including myself, jumped onto the value proposition of PhoneGap, which has iOS developers as an early target group. Without having to run the gauntlet of learning yet another language, I could be up and running in Xcode, building apps with only my knowledge of HTML5 as my guide.

Well, it's not quite that easy. Let's get started in building a basic PhoneGap solution in Xcode and look at how you can extend the functionality of the solution with plugins and custom Objective-C.

GETTING STARTED WITH CORDOVA FOR iOS

There are two places to start when developing with Cordova for iOS: the iOS Development Center and PhoneGap.com.

Let's start with iOS Development Center. First, do you have $99? Turns out that developing for Apple is not cheap. To even get in the door, you have to pay $99 per year to develop solutions for iOS. (Note: if you want to develop for Mac OS X, you need to pay another $99.)

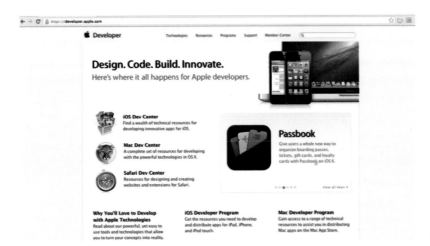

Figure 4.15: Apple's Developer website.

The fee covers your ability to sell your app. Unlike Android and many other competing mobile operating systems, Apple will allow you to sell your solutions only through the Apple-controlled iTunes App Store.

Access to Apple's Development Center gives you a whole host of services and tools, and there are lots of forums and sample code snippets for you to download. I will dig more deeply into all of these features when I cover native iOS development in chapter 7.

As soon as you have access to the Development Center, you will want to go ahead and download the latest version of Xcode from the OS X App Store. This does mean that you will need a Mac. Sorry, Windows guys, iOS apps *must* be developed on a Mac.

The next step you need to take is to go to PhoneGap.com and select the big, friendly button that says "Download PhoneGap." The file you will download is a ZIP file that contains all of the different development environments supported by PhoneGap, including iOS.

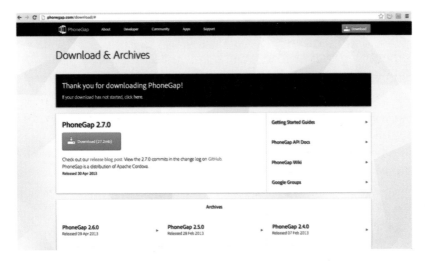

Figure 4.16: You can choose to download the current release of PhoneGap or any of the supported archive releases.

Now you are ready to create your first app.

CREATING YOUR FIRST APP

So, you have Xcode and the latest version of PhoneGap and have registered as an iOS developer. Time to build your first app.

1. Locate the Terminal app on your Mac. I keep Terminal on my Launcher across the bottom of the screen.

2. In your default "Documents" folder create a new folder and name it "Projects."

3. Extract the files from the PhoneGap ZIP file.

4. Open the subfolder named "lib."

5. Open the iOS subfolder.

6. Drag the "bin" subfolder onto the Terminal app. Terminal will open and will move the cursor to the "bin" subfolder.

7. In the Terminal, paste in the following command:

```
./create ~Documents/Cordova27/HelloWorld com.
CompanyName.HelloWorld HelloWorld
```

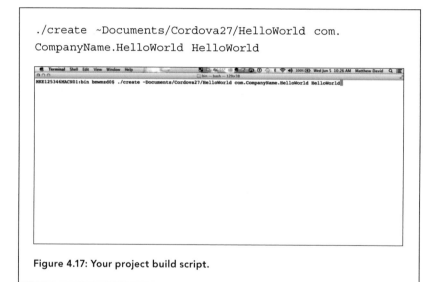

Figure 4.17: Your project build script.

8. The command line is in four pieces:
    a. The ./create is a command to build a new PhoneGapX code project.
    b. ~Documents/projects/HelloWorld, a path that you want to create the new project folder in.
    c. com.CompanyName.HelloWorld, the namespace convention Apple likes solutions to be listed in (your company website name followed by the app name).
    d. The name of the project (in this case, HelloWorld).

9. At this point you will want to press the "Return" key to run the program. What happens is that a new Xcode project is created.

Figure 4.18: A complete Xcode project is created by the build script. You can now develop your solution.

10. Your new project will be located in Documents—Projects—HelloWorld.

11. Open the new HelloWorld folder, and select the file names HelloWorld.xcodeproj.

12. Xcode will open. You are now running a new PhoneGap project.

13. Select the "Build" play button to compile the HelloWorld project in Xcode's iOS simulator. Your project will load as a native app on a simulated iPhone.

14. The content of the app can be changed in your project. Locate the folder titled WWW, and expand the folder. You will see a collection of HTML, JavaScript, CSS, and image files—yes, it is all Web content!

Figure 4.19: A successful build of your first PhoneGap native project.

15. Locate the file named "index.html," and select it. The following code will open in Xcode:

```html
<!DOCTYPE html>
<html>
<head>
<meta charset = "utf-8" />
<meta name = "format-detection" content = "telephone
= no" />
<meta name = "viewport" content = "user-scalable =
no, initial-scale = 1, maximum-scale = 1, minimum-
scale = 1, width = device-width, height = device-
height, target-densitydpi = device-dpi" />
<link rel = "stylesheet" type = "text/css" href =
"css/index.css" />
<title>Hello World</title>
</head>
<body>
<div class = "app">
<h1>Apache Cordova</h1>
<div id = "deviceready" class = "blink">
<p class = "event listening">Connecting to
Device</p>
<p class = "event received">Device is Ready</p>
</div>
</div>
<script type = "text/javascript" src =
"cordova-2.7.0.js"></script>
<script type = "text/javascript" src = "js/index.
js"></script>
<script type = "text/javascript">
app.initialize();
</script>
</body>
</html>
```

16.   Select the opening H1 element (<h1>Apache Cordova</h1>), and change the text to: <h1>You Rock!!</h1>.

17.   Save the file, and rebuild the project and run it in the iPhone simulator. Voilà! The app will load with your edits.

You have created and edited an app using just your knowledge of HTML. In chapter 7 I will step through the process you need to follow to submit your apps successfully to the iTunes App Store.

USING OBJECTIVE-C PLUGINS

Earlier in the chapter I mentioned that you could have the ability to extend the functionality of PhoneGap for iOS with support for native features. The PhoneGap team has built in a useful tool called plugins that allows you to do this easily.

The focus of plugins is to increase the functionality of your solution through leveraging native code. For instance, you can add a plugin that supports Facebook authentication. The code is written in Objective-C but with extensions in JavaScript that allow you to call the Facebook plugin from your HTML.

A collection of plugins can be found and downloaded at the following location: https:// github.com/phonegap/phonegap-plugins.

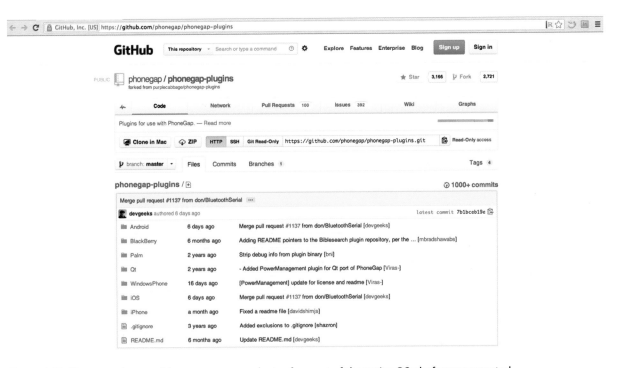

Figure 4.20: There are dozens of free open-source plugins for most of the native OS platforms supported through PhoneGap.

You will need a GitHub user ID and password to download a ZIP folder that contains native plugins for the following platforms:

- Android
- BlackBerry
- Palm
- QT
- Windows Phone
- iOS

Extract the files from the ZIP folder. Locate the iOS folder, and expand it.

Each plugin comes with a set of instructions that you must read, as each plugin is different. Some require additional resources (for instance, the Facebook plugin does require the official Facebook SDK for iOS), and all require custom features you will need to modify.

Each plugin has a "plugin" folder. Open your Xcode PhoneGap project, and you will see that there is a folder in your project with the name "plugin" into which you will need to drag and drop the "plugin" folder.

The instructions also explain how you can add the custom JavaScript into your default HTML page to connect to the app.

After that, it is simply a case of rebuilding your solution with the new custom plugin.

PLUGIN RESOURCES

Adobe is keeping an increasing list of articles that explain how to use a broad array of plugins at this great blog site: http://phonegap.com/blog/tag/plugin/.

AUTHORING YOUR OWN PLUGINS

Plugins are created with Objective-C and Xcode files for iOS solutions. If you have these skills, then you can create your own plugins. Adobe provides a comprehensive guide that steps you through the process at this Web address: http://docs.phonegap.com/en/2.7.0/guide_plugin-development_index.md.html.

WHEN HTML5 IS NOT ENOUGH—EXTENDING YOUR XCODE PROJECT WITH OBJECTIVE-C

The great thing when you work with PhoneGap is that you do not have to stay inside your HTML if you want to extend your project. To give you an idea of how easy it is to go beyond HTML, I am going to step you through the process of including Apple's proprietary iAd advertising tool in your app.

WHY INCLUDE ADVERTISING IN YOUR APPS?

At some point, you will want to get paid for the apps you create. Including advertising is a way to keep your app free but with the option of customers clicking on ads to give you revenue. Apple has introduced its own advertising solution, called iAds. OK, the caveat: iAds works only on iOS. No Android support here. The good news is that iAds are tied to your iTunes Developers Account, and it is easy to get paid. Indeed, you get paid every month by including iAds in your apps.

1. Open your PhoneGap iOS project in Xcode.

2. From the Build Phases tab, select the + in the Link Binary with Libraries.

3. Add the iAd.framework library.

4. Expand the "Classes" subfolder, and select the MainViewController.h file.

5. Add the following Objective-C:

```
#import <iAd/iAd.h>

@interface MainViewController : CDVViewController
{ ADBannerView *adView; }
```

6. Now locate and select on the MainViewController.m file.

7. Select the viewDidUnload method and add the following:

```
[adView release];
```

8. Locate the method named webViewDidFinishLoad and add:

```
adView = [[ADBannerView alloc]
initWithFrame:CGRectZero];

if([UIApplication sharedApplication].
statusBarOrientation = =
UIInterfaceOrientationPortrait || [UIApplication
sharedApplication].statusBarOrientation = =
UIInterfaceOrientationPortraitUpsideDown) { adView.
currentContentSizeIdentifier =
ADBannerContentSizeIdentifierPortrait;}else { adView.
```

```
currentContentSizeIdentifier =
ADBannerContentSizeIdentifierLandscape;}adView.
currentContentSizeIdentifier = ADBannerContentSizeIden
tifierPortrait;CGRect adFrame = adView.frame;adFrame.
origin.y = self.view.frame.size.height-adView.frame.
size.height;adView.frame = adFrame;[self.view
addSubview:adView];
```

9. Check your code to see if you have a method called willAnimateRotationToInterfaceOrientation to ensure that iAds always show on the top of the screen as you rotate your device. If you do not have the method, then add the following:

```
(void)willAnimateRotationToInterfaceOrientation:(UIInte
rfaceOrientation)newInterfaceOrientation
duration:(NSTimeInterval)duration {

BOOL hide = (newInterfaceOrientation = =
UIInterfaceOrientationLandscapeLeft||
newInterfaceOrientation = =
UIInterfaceOrientationLandscapeRight);

[[UIApplication sharedApplication] setStatusBarHidden:
hidewithAnimation:UIStatusBarAnimationNone]; CGRect
mainFrame = [[UIScreen mainScreen] applicationFrame];
[self.view setFrame:mainFrame];

if (newInterfaceOrientation ! =
UIInterfaceOrientationLandscapeLeft &&
newInterfaceOrientation ! =
UIInterfaceOrientationLandscapeRight) {

adView.currentContentSizeIdentifier =
ADBannerContentSizeIdentifierPortrait;

[self.view bringSubviewToFront:adView];

adView.frame = CGRectMake(0.0, self.view.frame.size.
height -adView.frame.size.height, adView.frame.size.
width, adView.frame.size.height);

}
else {

adView.currentContentSizeIdentifier =
ADBannerContentSizeIdentifierLandscape;

[self.view bringSubviewToFront:adView];
```

```
adView.frame = CGRectMake(0.0, self.view.frame.size.
width -adView.frame.size.height, adView.frame.size.
width, adView.frame.size.height);

}

}
```

10.   Save your work, and build your project. You will now see a test iAd displayed on your device.

This was a simple example, but you can see how you can build the majority of your project in HTML/JavaScript and leverage the extensive nature of PhoneGap inside Xcode to extend your project with native Objective-C.

Using Cordova to Create Apps for Other Platforms

What I like as you work with Cordova is that the framework is very similar from one development environment to another. The previous section was a deep dive into PhoneGap for iOS, but the principles for building solutions for iOS apply to all platforms: you start with HTML5, leverage plugins, and extend with native code. The following link will take you to PhoneGap's site, which steps you through what you need to know in building apps for Android, BlackBerry, iOS, Symbian, WebOS, Windows Phone 7, Windows Phone 8, Windows 8, Bada, and Tizen: http://docs.phonegap.com/en/2.7.0/guide_getting-gtarted_index. md.html#Getting%20Started%20Guides.

Using one framework and your knowledge of HTML, you can create native solutions for all leading platforms. In addition, it is clear that the open-source nature of Cordova provides an opportunity for additional platforms to create solutions for new mobile operating systems such as FireFox OS or Ubuntu.

Leveraging PhoneGapBuild

Earlier in the chapter I listed the many different platforms you can build solutions to using PhoneGap. But there is a problem. Many of the platforms require that you use a specific operating system to build the app on. For instance, iPhone apps must be built using a Mac. Windows apps must be built on Windows.

What if you cannot afford all of these devices?

This is where Adobe's cloud version of PhoneGap comes into play. The functionality of PhoneGapBuild is that it gives you the option to build solutions for many platforms without requiring any of those platforms.

YOUR BUILD SERVER IN THE CLOUD

The focus of PhoneGapBuild is to give you the tools you need to be able to create cross-platform solutions without requiring a build environment. With Build you can create solutions for iPhone, iPad, Android, Windows Phone, and BlackBerry.

There are some challenges. You will need to have device certificates for each platform (check out the great Build help docs), and you will not be able to create native code. But for smaller projects this is really OK.

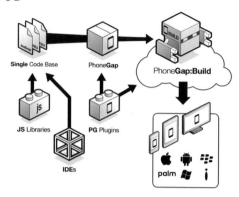

Figure 4.21: Build creates your apps in the cloud.

GETTING STARTED WITH PHONEGAPBUILD

The first step in creating your Build project is to create a Web app. Build requires that you include an instruction file within your Web app site at the root of the site. The file is called "Config.xml" and contains instructions for compiling your app.

Config.xml is an XML file and is very easy to set up. The first section of the XML document contains the following:

```
<?xml version = "1.0" encoding = "UTF-8"?>
<widget xmlns = "www.w3.org/ns/widgets"
xmlns:gap = "http://phonegap.com/ns/1.0"
id = "com.madlearning.sample"
version = "1.0.0">
```

The opening section declares the ID and version number for your app. The following sections are instructions in XML format. The following is the name of your app:

```
<name>Sample</name>
```

The next section is a description for what your app does:

```
<description>
```

This is just a sample application.

```
</description>
```

Next comes information about the author of the app:

```
<author href = "http://madlearning.com" email = "mdavid@
madlearning.com">
MAD Learning
</author>
```

The next section identifies all of the different icons needed for different platforms. The following is an iPhone icon for iOS:

```
<icon src = "res/icon/cordova _ ios _ 144.png" width = "144"
height = "144" gap:platform = "ios" />
```

The following is the information needed for an Android icon:

```
<icon src = "res/icon/cordova _ android _ 96.png" width = "96"
height = "96" gap:platform = "android" />
```

And, finally, a BlackBerry icon:

```
<icon src = "res/icon/cordova _ bb _ 80.png" width = "80"
height = "80" gap:platform = "blackberry" />
```

The following section lists the different types of splash screen as the app loads. The first is for Android:

```
<gap:splash src = "res/screen/android _ hdpi _ landscape.png"
width = "800" height = "480" gap:platform = "android" />
<gap:splash src = "res/screen/android _ hdpi _ portrait.png"
width = "480" height = "800" gap:platform = "android" />
<gap:splash src = "res/screen/android _ ldpi _ landscape.png"
width = "320" height = "200" gap:platform = "android" />
<gap:splash src = "res/screen/android _ ldpi _ portrait.png"
width = "200" height = "320" gap:platform = "android" />
<gap:splash src = "res/screen/android _ mdpi _ landscape.png"
width = "480" height = "320" gap:platform = "android" />
```

```
<gap:splash src = "res/screen/android_mdpi_portrait.png"
width = "320" height = "480" gap:platform = "android" />
<gap:splash src = "res/screen/android_xhdpi_landscape.png"
width = "1280" height = "720" gap:platform = "android" />
<gap:splash src = "res/screen/android_xhdpi_portrait.png"
width = "720" height = "1280" gap:platform = "android" />
```

The splash screen for BlackBerry:

```
<gap:splash src = "res/screen/blackberry_transparent_300.
png" width = "300" height = "300" gap:platform =
"blackberry" />
<gap:splash src = "res/screen/blackberry_transparent_400.
png" width = "200" height = "200" gap:platform =
"blackberry" />
```

The splash screen for iPad:

```
<gap:splash src = "res/screen/ipad_landscape.png" width =
"1024" height = "748" gap:platform = "ios" />
<gap:splash src = "res/screen/ipad_portrait.png" width =
"768" height = "1004" gap:platform = "ios" />
<gap:splash src = "res/screen/ipad_retina_landscape.png"
width = "2048" height = "1496" gap:platform = "ios" />
<gap:splash src = "res/screen/ipad_retina_portrait.png"
width = "1536" height = "2008" gap:platform = "ios" />
```

The iPhone splash screen:

```
<gap:splash src = "res/screen/iphone_landscape.png" width =
"480" height = "320" gap:platform = "ios" />
<gap:splash src = "res/screen/iphone_portrait.png" width =
"320" height = "480" gap:platform = "ios" />
<gap:splash src = "res/screen/iphone_retina_landscape.png"
width = "960" height = "640" gap:platform = "ios" />
<gap:splash src = "res/screen/iphone_retina_portrait.png"
width = "640" height = "960" gap:platform = "ios" />
```

The Windows Phone splash screen:

```
<gap:splash src = "res/screen/windows_phone_portrait.jpg"
width = "480" height = "800" gap:platform = "winphone" />
```

The following information lists which version of PhoneGap you should be using, whether or not orientation is fixed, and whether the app should be full screen (hiding the status bar):

```
<feature name = "http://api.phonegap.com/1.0/device" />

<preference name = "phonegap-version" value = "2.5.0" />

<preference name = "orientation" value = "default" />

<preference name = "target-device" value = "universal" />

<preference name = "fullscreen" value = "false" />

</widget>
```

There are many more settings you can configure, but these are the default settings you should be sure to use.

When you have all the files, you will want to compress all of the files into a ZIP folder and go to Build.PhoneGap.com. To access Build you will need an Adobe Creative Cloud license.

Log in with your Creative Cloud user ID and password.

On the first screen you will see a button that says + New App. Select the button and upload your ZIP file.

Figure 4.22: No need to have multiple build servers when you can use PhoneGapBuild to do all the work for you.

OVER-THE-AIR UPDATES WITH HYDRATION

A unique feature of PhoneGapBuild is a tool that allows you to perform over-the-air updates of yours apps. The feature is called Hydration and is activated by selecting a checkbox. After you have selected your checkbox, all future updates you make to the source code will be automatically pushed to any client you support.

Each build platform has unique app certification. Please read the small print for each one.

Each build will automatically run and show you whether or not your build was successful. If the build is good, then you will see a file you can download to upload to a public app store. A red button with a warning sign will appear if the build does fail. Select the warning sign to read why the build failed and how you can fix the problem.

So, without knowing any native code, you can build apps for all of the major platforms. Does is get much easier than that?

What PhoneGap Cannot Do

This chapter has covered a lot, and a lot has been said about how amazing PhoneGap is. With just a little HTML5 knowledge, you can rule the world. Or can you? This is the challenge you will need to wrestle with as you work with PhoneGap: when HTML5 is good but your client is expecting something great. Performance, custom code, and access to the latest feature in an OS are barriers for PhoneGap. Yes, it is a great tool for small projects, proofs of concept, and certain applications, but do not mistake it for a silver bullet that eliminates the need to learn native languages.

At some point your apps will get so big that the balance between keeping the center of the app in HTML5 and augmenting it with native code (Objective-C, Java, or C#) does not carry weight. This is why Facebook's native app famously threw away support for HTML5 in favor of a system that was 100 percent native.

So, the final word on PhoneGap is this: use it as a solution where appropriate. PhoneGap will serve you well, but you need to know a collection of tools to be a great mobile app designer.

CHAPTER 5

Leveraging ActionScript to Build Native Apps

Adobe is going to great lengths to make it very easy for you to develop mobile applications for iOS and Android. You have two choices: HTML or ActionScript. PhoneGap and Dreamweaver are the tools you will use if you want to leverage HTML, JavaScript, and CSS to create your apps. If, however, you are like millions of designers who have years of experience developing Flash solutions, then you will be relieved to know that your investment in ActionScript has not gone to waste.

The good news is that you can build Android and iOS apps (for phones and tablets) right out of the gate. There is no need to install extra features, as you used to have to do. All you need to do is download Flash CC and get started.

INSTALLING YOUR AIR APPLICATION ONTO AN ANDROID OS

Flash uses Adobe Integrated Runtime (AIR) to create your applications for Android. By default, AIR is not installed on the Android phone. This does not stop you from installing your new app; it will simply stop you from running it.

Fortunately, AIR is freely available in the Google Marketplace. If your phone does not have AIR installed, you will be prompted to download and install it from the Marketplace. There are no complex hoops to jump through. If you have installed one app, you know how to install AIR and enjoy all of the Flash apps in the Marketplace.

INSTALLING YOUR AIR APPLICATION ONTO iOS

Not so long ago, it was a pain to install iOS-created apps from Flash onto your iPhone or iPad. Now, all you need to do to test your apps is to connect your iOS device via USB to your computer. The latest release of Flash CC pushes out test apps directly onto your device.

Building Your First Application Using Flash CC

The goal for your first application is a simple one: to get a basic Flash movie running successfully on your phone. For this experiment I recommend that you use an Android phone, as it is easier to connect Flash CC to your Android OS device.

The following steps will take you through the whole process. At the end of the process you will have created your first native Android application using Flash tools. The next section explains how to install the Android application onto your device.

1. Begin by opening Flash CC. Select File ◊ New to open the new file window.

2. Select "Template" from the top button of the new file window.

3. Choose "Air for Android" from the left category window. On the right side you will see "800×480Android." Select OK.

Figure 5.1: Flash CC has a template for Android applications.

To keep things simple, we are going to create all we need for a simple test. Save your file to your hard drive. Name the file "FirstApp.fla."

4. On the Stage use the text tool to draw a text region.

5. Set the font to "_sans."

6. Change the font size to "20."

7. With the text field still selected, change the text type to "Dynamic Text."

8. Give the text field an ID of "txt."

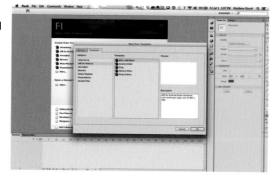

Figure 5.2: Currently there is only an AIR for Android template, but you can create your own for tablet devices.

9. Open the Actions window (or press F9), and add the following ActionScript. The goal for this is to show you that the ActionScript you have been using all along will work. Enter the following ActionScript:

```
txt.text = "hello, world";
```

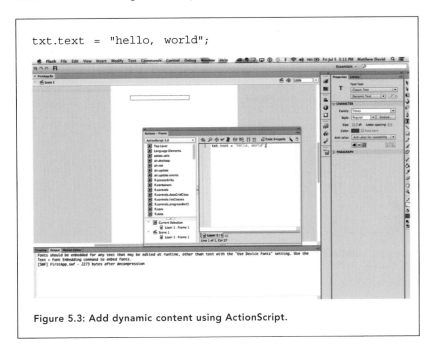

Figure 5.3: Add dynamic content using ActionScript.

10. At this point you can test your Flash movie by pressing CTRL+ENTER on Windows or Command+Enter on Mac. The movie should show you the text "hello, world" on your screen.

11. The next steps are to convert the Flash movie into an Android application.

12. Select the Stage, and choose the Properties panel. In the Profile section you will see "Air for Android Settings." Select the "Edit . . ." button.

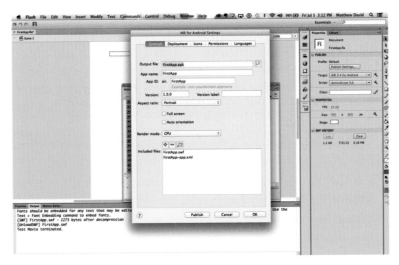

Figure 5.4: The Android application is created using special publish settings.

13.   The Application and Installer options window will open. Across the top of the window you will see three buttons that toggle three different settings windows. The buttons are General, Deployment, and Icons.

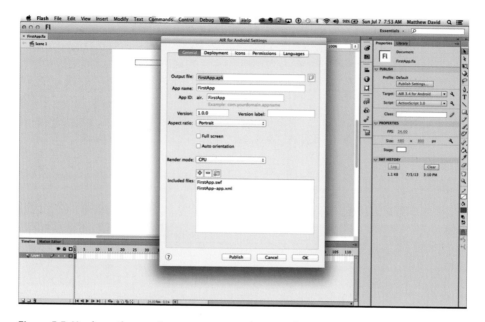

Figure 5.5: You have three main screens you need to modify to create your Android apps.

14.   The General button shows you the following settings:
      a.   Output file
      b.   App Name
      c.   App ID
      d.   Version
      e.   Aspect Ratio
      f.   Full Screen
      g.   Auto Orientation
      h.   Included Files

15.   The Output file is the location of the final file that will be installed on your Android device. The file format for Android apps is APK. For this example you can keep the default file name. It should be called "FirstApp.apk" and will save to the same folder as your Flash FLA file.

16.   The App Name is the name of the app as it will appear on the Android phone. The default is to use the name of the FLA file. Change the name to read "My First App."

17.   The App ID is used when you publish your app to the Marketplace. For now you can keep the default "FirstApp."

18.   The version number allows you to add a version number to your Android app. It is up to you how you want to number your versions.

19. The Aspect Ratio forces the default presentation of your Android app into either Landscape or Portrait. For now, keep the Aspect Ratio to Portrait. Later, when you develop your first games, you will learn how to design for Landscape Aspect Ratio.

20. Select the checkbox for Full Screen. The Full Screen setting forces the application to use up the whole of the screen and hide the status bar on the Android phone.

21. Do not select the Auto Orientation checkbox. Auto Orientation will allow the app to rotate as you rotate your phone.

22. The "Included Files" section allows you to add additional files into your final APK package. This can include files such as video, audio, and other SWF movies. You do not need to worry about that at this time.

23. Now, select the "Deployment" button to go to the Deployment screen.

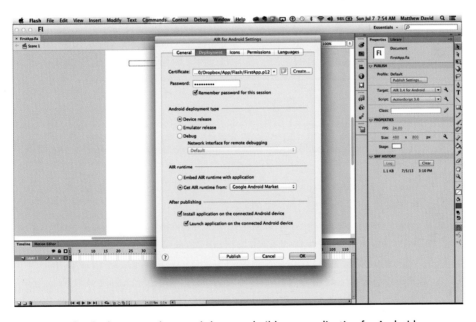

Figure 5.6: The deployment tab controls how you build your application for Android.

24. Each AIR app you build for Android requires a certificate. For development purposes you can use the same certificate over and over. Let's create a Developer Certificate.

25. Select the "Create" button. A new screen will open asking you for additional information for the certificate.

26. For Publisher Name, Organization Unit, and Organization Name, insert "Self." This is not a magical term; you can really enter anything you want.

27. Select the country from the drop-down menu.

28. Enter a password. Make sure you remember the password as you will need to use for future applications.

29.   You can use the default 1024-RSA certificate strength.

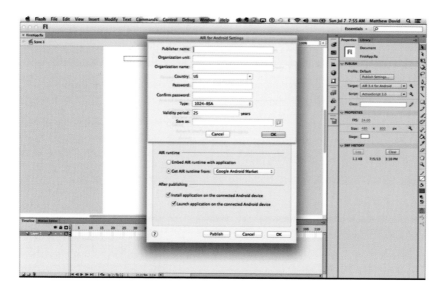

Figure 5.7: A certificate is valid for 25 years.

30.   The default validity period is 25 years. That should be good enough for what we are doing.

31.   Select the folder where you would like to store the certificate. The certificate will default to the file name "mycert.p12."

32.   Select OK. A window stating a "Self signed certificate has been generated" will pop up. Select OK.

33.   You will go back to the Deployment window. Enter your password. Choose the "Remember password for this session" checkbox. While you have this FLA file open, you will not need to keep re-adding the password each time you compile the file.

34.   The Android Deployment Type option allows you to choose "Device debugging" or "Release." For now, select the "Device debugging" option.

35.   Flash can install the final APK file directly onto your Android device for you. You will need to have the downloaded Android SDK. The "After Publish" section will install the Application on your device, but you need to have the Android SDK ADB tools. You can find the ADP tools within the Android SDK's "tools" folder.

36.   Don't worry about icons at this time.

37.   Select the "Publish" button.

38.   The app is small and should take only about 15 seconds to publish. You have now created an APK file, and, if you selected the install options, you now have your first Android app running on your phone. How cool is that? Knuckle punch!

At this point you have your first application running on your Android phone. The good news is that now that you have one application running you do not need to go through the hard work of installing JRE, Android SDK, AIR for Android, or a developer's certificate again. You have done the hard work. Now you can focus on creating great AIR solutions with Flash CC for the Android platform.

Designing Apps for the Android OS

Touch-based interfaces, whether they are Android, WebOS, or iOS, all follow similar rules in design. These are the design aspects you need to keep in mind:

- Interacting with a mobile phone
- Multitouch and gestures

Keeping these four concepts in mind will cause you to approach the design of your application differently but with a focus on the final product.

INTERACTING WITH YOUR MOBILE PHONE

The primary input tool for your mobile phone is your finger—fat or thin, your digit is the thing. Again, like the move to smaller screens, using your finger as an input device requires a new design technique that changes how you design your apps.

Take a look at your finger tips. You may have long nails, short and stubbies, long and graceful fingers or, like mine, finger tips that would look better on a construction worker. In other words, we do not all have similarly shaped fingers. In contrast, your mouse for your computer is similar to any other mouse you can buy. The precision of a mouse is consistent from one mouse to the next. The same cannot be said of your fingers.

Even though we are talking about Android development, you will want to check out Apple's Human Interface guidelines for iPhone development (http://developer.apple.com/iphone/library/documentation/userexperience/conceptual/mobilehig/). Apple details the way you should approach your mobile development. It is a good read and can be practically applied to your Android development.

When it comes to input on your screen using your fingers, you will want to think big. Approximately 44 px by 44 px big. A 1 cm square is approximately the smallest area your finger can accurately select on a touch screen.

Figure 5.8: Using your finger to interact with your phone.

On the whole, single taps will be the most common way you interact with your Android phone. Drive your design for this type of interaction. A good example of where you will want to change typical design practices is checkboxes in a form. Typical checkboxes usually take up only about 5 × 5 px of space—too small for fingers.

Always be honest with yourself as you test your apps on your test devices. As you tapped the screen, did it really feel comfortable, or did you have to perform some yoga trick with your pinkie finger to complete the interaction?

Later, in the ActionScript section of the chapter, we will dig deeply into optimizing finger interactions with your Android phone.

WORKING WITH GESTURES AND MULTITOUCH

Earlier we covered the limitations of using your fingers to touch your device. Let's switch things around and cover what the advantages are in using your fingers. With Flash AIR on your Android phone, you can now do the following:

- Multifinger input
- Gestures

Ever thought it would be great to have two or more mouse events on a screen at once? AIR for Android supports five simultaneous touch points on a screen. Want to create a piano app that requires that two or more keys be tapped at the same time? Multitouch will come to your rescue.

The way multitouch works is defining what the first touch point on the screen is and then what can be accepted as multiple touch points beyond that.

Figure 5.9: Multitouch can be used from Flash on many different devices, such as a Windows Surface.

Apple brought to the world's attention the ability to perform gestures. A gesture is an action that you do with your fingers on the screen. The most common gesture is a "swipe" gesture. For instance, you may be looking at the e-mails in your inbox and want to delete an entry. Instead of tapping on the e-mail and then selecting a delete button, you can use the "swipe" gesture to drive a different type of interaction and result. Swiping across the e-mail entry can trigger an action that deletes your message.

There are many different gestures you can use in your Flash design. They include but are not limited to the following:

- Pan
- Rotate
- Zoom

- Press and Tap

- Double Tap

The Pan gesture is similar to holding the mouse down on your computer and dragging something across the screen. With the Pan gesture you tab and hold your finger down on the screen and drag an object around.

The Rotate gesture is performed with two fingers. A good example to demonstrate how this works is to use an image gallery metaphor. To rotate an image, you use two fingers to tap on an image and then keep pressing down on it. Twisting your wrist, you can rotate the image. This can be accomplished with any movie clip you create in Flash.

Zoom is very similar to Rotate. With Zoom you select an object on a screen with your finger and thumb and then squeeze in or spread out the distance between your thumb and finger. The object will zoom in or out.

A more complex gesture is Press and Tap. With Press and Tap you select an object on the screen and then have the ability to tap on a second object. This allows for applications where two or more people are interacting on the screen at once. While this may be difficult on small phone screens, larger screens for Android-powered tablets easily have enough screen for multiperson solutions, such as split-screen games.

The final commonly used gesture is double tap. Double tap can be thought of the same as double-click with your mouse. On a Windows desktop you can open an application by double-clicking the icon. Similarly, if you use one finger and double tap on the screen, you can cause an action to happen in your Flash movie.

Multitouch and gestures do require use of ActionScript. We will be getting to that very soon. For now, bask in the knowledge that you can add an input dimension to your work that, until recently, was not physically possible.

All of the techniques you learn and use in this chapter will work in Android and iOS. But what you learn does not end here. The number of devices arriving on the market that support touch-driven interfaces is increasing (thanks largely to the success of Apple's iPhone and iPad). It is common now to find monitors that are touch driven, laptops that support gestures, and a lot of other mobile devices. The touch and gesture techniques you learn in this chapter can be used with AIR and Flash applications on the following:

- Windows 7, 8, and 8.1

- Mac OS X (10.5.3 and later)

- Windows Mobile 6.7

- Windows Series 7 and 8 phones

- Android

- BlackBerry 6 and later

Multitouch events are driven using ActionScript. This is going to be covered in more detail later in the chapter, but, to give you an example of how easy it is to use, let's step through two different scripts: single touch and zoom.

Touches, taps, and gestures are all managed through listeners in your ActionScript. A listener's job is to react when an event happens. The most common type of listener reacts when you click on a movie clip with your mouse. The following ActionScript will rotate a movie clip called "myMovieClip" on the Stage by 25 degrees each time you click it:

```
myMovieClip.addEventListener(MouseEvent.CLICK, onClick);

function onClick(event:MouseEvent):void

{

myMovieClip.rotation + = 25;

}
```

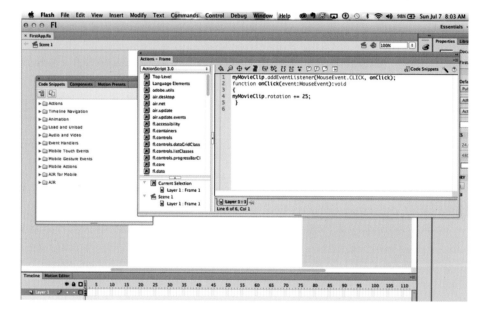

Figure 5.10: Traditional ActionScript mouse events can be used as single-tap events on the Android OS.

The first line of the ActionScript is waiting for the mouse to click on the movie clip. The ActionScript will do nothing until the listener is triggered.

When a mouse clicks on the movie clip, a function is then called into play. In this case, the function is called "onClick."

The second line is the "onClick" function. The function is driven by a mouse event and runs the ActionScript in the curly brackets. In this case there is only one line of code in the curly brackets—a script that causes the myMovieClip to rotate 25 degrees.

This is a listener that you may well have used many times in other Flash movies. The good news is that this same script will work for a single tap from a finger on an Android phone. Without having to rework all your existing movies, you can use mouse events such as "Click" as a quick substitute for simple touch events.

Gestures are more complex but follow the same pattern as the mouse click listener just described. Here is an ActionScript that allows you to take a movie clip and zoom in on it. Remember, the zoom requires two digits (normally a thumb and index finger) to select an object and manipulate it.

```
Multitouch.inputMode = MultitouchInputMode.GESTURE;

myMovieClip.addEventListener(TransformGestureEvent.GESTURE_
ZOOM, onZoom);

function onZoom(e:TransformGestureEvent):void

{

var myMovieClip:Sprite = e.target as Sprite;

myMovieClip.scaleX * = e.scaleX;

myMovieClip.scaleY * = e.scaleY;

}
```

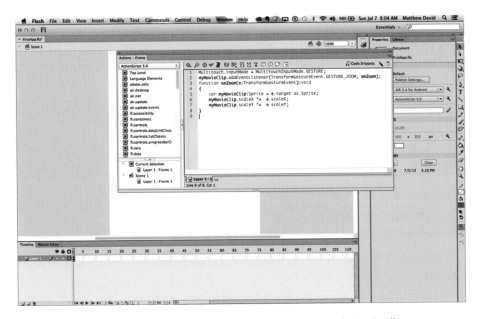

Figure 5.11: MouseDown events (such as selecting with your mouse button) will convert automatically to single-tap events.

The block of ActionScript presented here can be broken in two sections:

1.   Event Listener

2.   Zoom Function

The first line declares that a gesture is going to be applied to the listeners. The second and third lines declare the gesture you are going to use: zoom (GESTURE_ZOOM).

The function, lines 4–9, controls the zoom gesture. The function controls the Z and Y scale properties. Overall, not too much more complicated than a mouse click of single-tap event.

Controlling the Use of Fonts

The use of fonts is something I am very particular about. Most of the content you interact with is text driven. This is true for mobile devices. Fortunately, Flash gives you three ways to use text in your design. The three techniques you can use are:

- Local fonts
- Embedding fonts with Classic Fonts
- Embedding fonts with the new Text Layout Framework

It can be difficult to control what fonts a user has installed on his device. If you are a Web developer, you are already aware of this restriction in your Web design. Unless you embed a font into your Flash movie, you are restricted to a small number of fonts. They are:

- _sans
- _serif
- Clockopia.ttf
- DroidSerif-Bold.ttf
- DroidSans-Bold.ttf
- DroidSerif-BoldItalic.ttf
- DroidSans.ttf
- DroidSerif-Italic.ttf
- DroidSansFallback.ttf
- DroidSerif-Regular.ttf
- DroidSansMono.ttf

It is advisable to use local device fonts when you have text fields that require a user to input text. The fonts simply render faster.

To embed a font feature, you have to follow these steps:

1.   Using the Text tool in Flash, draw a text object on the screen.

2.   Select the new text area. In the Properties panel, select Classic Text from the frameworks drop-down.

3. In the Character section, change the font to one you would like to embed. Select the Embed button.

4. When you select the Embed button, the Font Embedding window opens. The Font Embedding window gives you options for controlling which font properties you want to embed.

5. Give your font a name. The name will be a reference you can use in your Library if you want to embed your font in other text areas.

6. In the Character Range, choose which parts of the font you want to embed. You will see that Flash gives you different options, such as embedding the whole font, just lower-case letters, upper-case letters, numerals, and punctuation. Embedding a font increases the file size of your final movie. These different options will help you streamline how much of a font you really need. Select the OK button.

7. Save your work.

At this point you have done all the work needed to successfully embed a font. It is important to note that when you embed a font, you are embedding only the outline of the font, not the whole font. This protects the original copyright owner of the font.

ADOBE DROPS SUPPORT FOR TLF

Flash CS 5.5 released a new way to control fonts in AIR/Flash solutions called TLF. It is important to know that Flash CC has dropped support for TLF. So, if you are one of the few who transitioned, you have a bit of work to convert your apps back to the Classic Font management. If, like most of us, you did not use TLF, your solutions are OK.

Working with Image Files in Your Apps

When you are working in Flash, there are two basic file types you can use: bitmap and vector.

Bitmap images are constructed pixel by pixel. You typically find bitmap images used on the Web and in your digital camera. A bitmap image is a photo-realistic format. Popular bitmap image formats include JPEG and PNG.

Vector images are created mathematically. If you use Adobe's Illustrator, then you are used to working with the world's most popular vector drawing tool. Vector images possess a feature that allows them to change to any scale and still retain crisp lines. This is because the images are created mathematically, unlike bitmap images, which are created to specific pixel density and file size.

The traditional approach to creating Flash animation is to do so with vector images. The elegant method for easily scaling the final vector drawings to fit any screen size makes this an ideal solution when you do not know the size of the final screen.

You might think, then, that vector images are the way to go when you do not know the screen size for your Android phone. It is not. When it comes to images on your Android phone, stick to bitmap images wherever you can.

In fact, what you want to do, where possible, is to not have any images at all. By this, I mean that you will want to use the default Stage color for your background wherever possible. This will help speed up your application.

The reason for using bitmap images over vector images comes back to the limited power of the processor on your phone. Vector images require more processing power than bitmap images. Managing power is a key element of your development.

In addition, not all bitmap image formats are created equal. The ideal bitmap image format is Progressive Network Graphic (PNG), an image format used by many Web developers. The PNG file format has all of the benefits of other popular bitmap formats such as JPG, but PNG files are photo realistic and come with these other advantages:

- PNG images support Alpha levels for transparency.
- Android uses special accelerators for PNG images in its GPU.
- PNG images can be loaded on demand into Flash.

It is a big switch for Flash designers. In the past you have been told to use vector images. Now, you are being told to use PNG bitmap images. The end result is using the right image and format for your target end device.

Leveraging Custom Device Hardware Calls with ActionScript

A smartphone gives you a lot of additional controls, such as multitouch, gestures, Accelerometer, and Geolocation. In this section you will learn how you can tap into the hardware-specific extensions with ActionScript to add a rich level of control to your apps.

Specifically, we are going to review the following:

- Gestures
- Orientation
- Geolocation
- Loading data into Flash
- Microphone
- Camera/Video

The AIR platform is maturing at a rapid clip. The current release is 3.7 and comes loaded with great features (most of which will require their own book to cover).

With that said, the mobile features covered in this chapter will get you up and running very quickly.

The first set of changes you will make will allow you to load data from remote sites onto your Android device. After this, you will start to interact directly with the hardware on the phone itself.

LEVERAGING NATIVE OBJECTIVE-C OR C++ IN YOUR FLASH APPS

AIR now supports ActionScript Native Extensions (ANE) that you can use to connect to native APIs on a device such as GameCenter for the iPhone. For more information on ANE go here: www.adobe.com/devnet/air/native-extensions-for-air.html.

Using Gestures in Your Apps

Adobe includes many programmable interfaces you can use through ActionScript. Multitouch is a feature you may use in many of your applications. This section explains how multitouch is programmed into your apps.

- Using your finger instead of a mouse to interact with applications
- Using two or more fingers in your App

In many ways, it is the use of your fingers that makes touch so compelling on iOS and Android devices. But there are some caveats you need to keep in mind as your little digits tap on your OLED screen.

- Not all touch screens are the same—the king of sensitivity is the iPhone. No matter where you touch the screen, you will get the desire response. In contrast, the original Motorola Droid was a big disappointment for sensitivity. You often find yourself repeatedly tapping the same area before you get the desire responses (Note: the new Droid Incredible is much better).

- Your fingers are not as delicate as a mouse—the reality is that a mouse or stylus is a much more accurate pointing device than your fingers. Keep this in mind as you design your apps.

- Fingers tend to be big—Apple states in its Human Design guidelines that you should allow for 44 px × 44 px (height and width) to accommodate the average finger

- Simultaneous tap—you can have up to eleven fingers tapping the screen simultaneously for iOS devices and five for Android. Not sure why it is eleven and not twelve or just ten, but I did not develop the code

Keep these four rules in mind as you use control content on the screen.

EXTENDING YOUR GESTURE CHOICES IN AS3

There are some great open-source libraries you can leverage with ActionScript 3 for apps. The following link is for a library you can use to extend the number of gestures you want to leverage in your solutions:

https://github.com/fljot/Gestouch.

USING A SINGLE FINGER TO INTERACT WITH CONTENT

There is a lot of hoopla about gestures and multitouch development. But we have been getting away with just a single tap of the mouse button for many years. You will also find that most of the time a single tap from one finger is really all you need. The great thing with using a single tap is that the event is exactly the same as a single mouse click. You use the MouseEvent.CLICK to trigger a single-tap interaction. Let's see this in action.

1. Open a new Flash Android or iPhone application.

2. On the Stage, draw a rectangle. Press the F8 button to convert the drawing into an object.

3. Name the instance of the rectangle on the "myObject."

4. Select frame one on the Timeline.

5. Open the Actions Panel. Paste the following ActionScript into the screen:

```
myObject.addEventListener(MouseEvent.CLICK, fl_Mouse
ClickHandler);

function fl_MouseClickHandler(event:MouseEvent):void

{

myObject.alpha * = 0.5;

}
```

6. The code just presented is essentially a simple listener that is looking for a mouse click. The good news is that a single click is the same action your finger is applying to the screen.

7. Test your movie in your Android or iOS device. What you will see is that, as you tap on the screen, the Alpha level of the rectangle of the screen will change.

Using the MouseClick event is a great trick when you want to quickly migrate code from a standard desktop app to a Web app. There is, however, a better way to do this using the TouchEvent listener.

Flash 10 introduced a slew of multitouch events you can use, the simplest of which is a single tap. The following code duplicates the exact same action seen using the MouseClick by using the TouchEvent class.

```
Multitouch.inputMode = MultitouchInputMode.TOUCH _ POINT;
myObject.addEventListener(TouchEvent.TOUCH _ TAP, fl _ Tap
Handler _ 2);
function fl _ TapHandler _ 2(event:TouchEvent):void
{
myObject.alpha * = 0.5;
}
```

What you will see as the main difference is that the TouchEvent is specifically looking for a single tap on the screen (the TOUCH_TAP event).

DRAGGING OBJECTS ACROSS THE STAGE

A common practice when you are building interactive applications is to drag objects across the Stage. This has been successfully done with the mouse for more than a decade in Flash. So, can you do the same with your finger?

The action you are looking to create is called a "gesture." You tap, hold, and drag an object across the Stage. That's it. The multitouch class used in the TouchEvent just described is once again leveraged to add this gesture.

A drag event is defined by two events: the place from which you drag your object and the place to which you drag the object. You do this in Flash by using two event listeners (one for the Begin Event and the second for the End Event) that trigger two separate functions.

For instance, you can use the same instance created I just described and add the following code:

```
Multitouch.inputMode = MultitouchInputMode.TOUCH _ POINT;
myObject.addEventListener(TouchEvent.TOUCH _ BEGIN, fl _ Touch
BeginHandler);
myObject.addEventListener(TouchEvent.TOUCH _ END, fl _ Touch
EndHandler);
var fl _ DragBounds:Rectangle = new Rectangle(0, 0, stage.
stageWidth, stage.stageHeight);
function fl _ TouchBeginHandler(event:TouchEvent):void
{
```

```
event.target.startTouchDrag(event.touchPointID, false,
fl _ DragBounds);
}
function fl _ TouchEndHandler(event:TouchEvent):void
{
event.target.stopTouchDrag(event.touchPointID);
}
```

The first line declares that you are using a new multitouch event. In this case, the event is called TOUCH_POINT. By declaring TOUCH_POINT you can now allow the object on the Stage to be dragged around.

The second line is the first event listener. In this case, the first event listener controls the start of the drag. You will see that the TOUCH_BEGIN event is paired with the function fl_TouchBeginHandler. The fl_TouchBeginHandler function is triggered on the fifth line. You will want to define where you can drag your movie clip in the TOUCH_BEGIN event. The fl_TouchBeginHandler function calls a variable on line five that affects how objects move on the screen.

Line screen is the TOUCH_END event, or what happens when you have dragged your object around the screen and are letting go. As with the first listener, the TOUCH_END listener is linked to a function. Here the function is stopping the drag action.

You can test this code in your movies to drag labeled objects around the Stage.

ADDING A LONG-PRESS EVENT TO YOUR CODE

What if you want to add a function such as holding a button down? There are many apps that are designed to measure how long you can hold a button on the screen. Fortunately, this is very easy to duplicate in Flash by mixing up your knowledge of ActionScript: using multitouch and timers.

A timer, as covered earlier, is an event that is controlled by time. In the following example you will add the code needed to increase the size of the main object on the Stage after the object has been tapped for one second.

1.   Let's just use the movie setup earlier. You should have shape on the screen with the ID "myObject."

2.   Go ahead and open the Actions panel. Begin by adding the ActionScript that will trigger a function when the movie clip is selected:

```
Multitouch.inputMode = MultitouchInputMode.TOUCH _ POINT;
myObject.addEventListener(TouchEvent.TOUCH _ BEGIN,
fl _ PressBeginHandler);
```

3.   The following is the function being called by the TOUCH_BEGIN event.

```
function fl _ PressBeginHandler(event:TouchEvent):void
{
fl _ PressTimer.start();
}
```

4.   The function is calling a variable called fl_PressTimer. This variable is associated with a timer listener. The following timer listener is set to a delay of 1,000 milliseconds. You will see that the listener calls a function named fl_PressTimerHandler that changes the size of the movie clip.

```
var fl _ PressTimer:Timer = new Timer(1000);
fl _ PressTimer.addEventListener(TimerEvent.TIMER,
fl _ PressTimerHandler);
function fl _ PressTimerHandler(event:TimerEvent):void
{
myObject.scaleX = 2;
myObject.scaleY = 2;
}
```

5.   The final step in your code is to add a second touch event that listens for when you lift your finger off the screen. The following does exactly that and runs a function that returns your movie clip to its original size.

```
myObject.addEventListener(TouchEvent.TOUCH _ END,
fl _ PressEndHandler);
function fl _ PressEndHandler(event:TouchEvent):void
{
fl _ PressTimer.stop();
myObject.scaleX = 1;
myObject.scaleY = 1;
}
```

6.   At this point, you save your file and publish to either your iPhone or your Android device.

As you can see, Adobe gives you many different ways to control a single finger's interaction on the screen.

Working with Gestures

The iPhone brought a new way of controlling your screen: gestures. A gesture is the use of two or more fingers simultaneously on the screen. Common gestures include:

- Two-finger tap
- Pinch and zoom
- Rotate
- Swipe

Each of these actions can be duplicated in Flash for use on your Android or iOS device.

ADDING TWO-FINGER TAP CONTROL

The two-finger tap is similar to a single-finger tap. Of course, the main difference is that you use two fingers. I know, give me a prize for pointing out the obvious. So, without much further ado, let's jump into the code.

As you expect by now, the multitouch class controls the event. The first line of code declares a new GESTURE event:

```
Multitouch.inputMode = MultitouchInputMode.GESTURE;
```

The second line of code states where the gesture is to be applied and what type of gesture it will be. In this case, the whole Stage is listening for the GESTURE_TWO_FINGER_TAP event.

```
stage.addEventListener(GestureEvent.GESTURE _ TWO _ FINGER _ TAP,
fl _ TwoFingerTapHandler);
```

An event is triggered when two fingers tap the screen:

```
function fl _ TwoFingerTapHandler(event:GestureEvent):void
{
myObject.scaleX * = 2;
myObject.scaleY * = 2;
}
```

You can swap out your code in the function for your own action.

That's it. As you can see, Flash has made it very easy for you to add a two-finger gesture.

ADDING PINCH AND ZOOM

Apple's inclusion of pinch and zoom has become almost a must-have for any photo album. Good thing you can do this in Flash.

As you would expect, you use a gesture to zoom an object on the Stage. Two fingers are required to pinch in and out. ActionScript refers to this as a TransformGestureEvent. The actual event is called GESTURE_ZOOM. You will see from the example here that the code is similar to that for a single-tap multitouch event with the exception of the event type in line two:

```
Multitouch.inputMode = MultitouchInputMode.GESTURE;

stage.addEventListener(TransformGestureEvent.GESTURE _ ZOOM,
fl _ ZoomHandler);

function fl _ ZoomHandler(event:TransformGestureEvent):void

{

myObject.scaleX * = event.scaleX;

myObject.scaleY * = event.scaleY;

}
```

You can test this on object on the Stage. Just change function to call the object you are manipulating.

ROTATING A MOVIE CLIP ON THE STAGE

This will come as no shock, but a rotate gesture is almost identical to a zoom gesture. As with the zoom gesture, you use two fingers. The difference is that your two fingers are anchor points around which you can rotate an object.

In the following code, the event you are looking for is TransformGestureEvent.GESTURE_ ROTATE. That's it. The function uses the rotation property in ActionScript to rotate the selected object.

```
Multitouch.inputMode = MultitouchInputMode.GESTURE;

myObject.addEventListener(TransformGestureEvent.GESTURE _
ROTATE, fl _ RotateHandler);

function fl _ RotateHandler(event:TransformGestureEvent):void

{

event.target.rotation + = event.rotation;

}
```

It is certainly a lot easier to add rotation in Flash than in Java for Android or Objective-C for iOS.

SWIPING OBJECTS ON THE SCREEN

In many ways, the most complex gesture you will accomplish on a mobile device involves swiping objects on the screen. The swipe gesture has the basic rule of dragging your finger across the screen. This is a common activity in data-driven applications on the iPhone.

What is really happening with a swipe event? When you swipe your finger on the screen, what you are sending to the device is an instruction to move select content on the screen to the left or right a specific number of pixels. In the following example you are going to move a movie clip to the left or right (depending how you swipe) by 40 pixels. You will also be able to swipe up and down moving the object 40 pixels.

The difficulty using the swipe gesture comes in controlling whether you swipe horizontally or vertically. You will define this in the gesture's event function, where you will look for either an X (horizontal) or Y (vertical) interaction.

1. The first line of code you need to add triggers the gesture.

```
Multitouch.inputMode = MultitouchInputMode.GESTURE;
```

2. The second line triggers the TransformGestureEvent for a GESTURE_SWIPE. The GESTURE_SWIPE is the event defined for a swipe. At this point, the ActionScript does not know the direction of the swipe.

```
stage.addEventListener (TransformGestureEvent.
GESTURE _ SWIPE, fl _ SwipeHandler);
```

3. The gesture's function is broken into two switch statements. The first statement examines whether the swipe action is left or right and then moves the object on the Stage accordingly.

```
function fl _ SwipeHandler(event:TransformGestureEvent):
void
{
switch(event.offsetX)
{
case 1:
{
myObject.x + = 40;
break;
}
case -1:
```

```
{
myObject.x - = 40;
break;
}
}
switch(event.offsetY)
{
case 1:
{
myObject.y + = 40;
break;
}
case -1:
{
myObject.y - = 40;
break;
}
}
}
```

4. The second switch statement examines whether the swipe is up or down.

5. Save your file and test.

As you can see, swiping does require additional code. With that said, it is not too complex.

ADDING TWO OR MORE GESTURES TOGETHER

Are you restricted to adding just one gesture to an object? No, you are not. The following script demonstrates how you can add a left/right swipe, rotate, and pinch/zoom gesture to the same object on the Stage:

```
Multitouch.inputMode = MultitouchInputMode.GESTURE;
stage.addEventListener (TransformGestureEvent.GESTURE _
SWIPE, fl _ SwipeHandler);
function fl _ SwipeHandler(event:TransformGestureEvent):void
{
switch(event.offsetX)
```

```
{
case 1:
{
myObject.x + = 40;
break;
}
case -1:
{
myObject.x - = 40;
break;
}
}
}
myObject.addEventListener(TransformGestureEvent.GESTURE_
ROTATE, fl_RotateHandler);
function fl_RotateHandler(event:TransformGestureEvent):void
{
event.target.rotation + = event.rotation;
}
stage.addEventListener(TransformGestureEvent.GESTURE_ZOOM,
fl_ZoomHandler);
function fl_ZoomHandler(event:TransformGestureEvent):void
{
myObject.scaleX * = event.scaleX;
myObject.scaleY * = event.scaleY;
}
```

Test this code to see it run on your device.

Gestures are huge part of your interactive development, whether it is for the iPhone, an Android device, or the BlackBerry.

Which Way Is Up? Controlling Orientation with the Android Accelerometer

The Android Accelerometer controls the orientation of the device. The same is true for iOS. In this section you will learn how you can interpret orientation through ActionScript to change the display for the correct screen position.

There are two ways in which you can control orientation in your Android apps.

1.   Publish Settings

2.   ActionScript

The easiest way to detect orientation is through the AIR Android Publish Settings. Select the Properties panel and choose AIR Android Settings. The Application and Installer Settings window will open. On the General tab you will see a check mark for Auto orientation. Select this check mark. Now, when you rotate the Android device, you will see your AIR app also rotate.

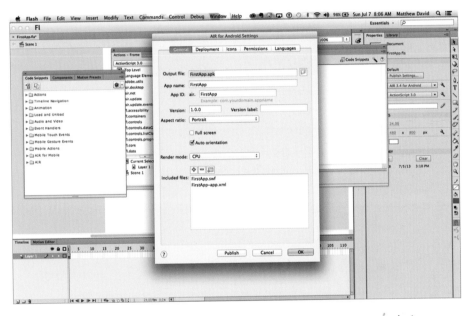

Figure 5.12: Select Auto Orientation to have your app rotate as you rotate your device.

This is easiest orientation tool, but it does not give you a lot of control. For this, you need to use ActionScript.

ADDING THE ACCELEROMETER TO YOUR APPS WITH ACTIONSCRIPT

With the release of the Flash Player 11 and Adobe Integrated Runtime (AIR) 3.7, the Flash team added several new core features. Access to the Accelerometer is one of these. The role of the Accelerometer is to detect when you move your phone. The Accelerometer is a listener that is triggered when it is used. The following example adds an Accelerometer listener to your iPhone app.

1.   Start by creating a new iPhone app and adding the necessary development properties in the iPhone settings.

2.   Add a dynamic text field to the Stage. Name the new text field "myTextField" in the Properties panel.

3. Create a new layer on the Timeline and name it "Actions." Select the "Actions" layer, and open the Actions panel. The step you need to take is to import the libraries for the Accelerometer to work correctly:

```
import flash.events.AccelerometerEvent
import flash.sensors.Accelerometer;
```

4. Now you need to create a new Accelerometer object:

```
var acc1:Accelerometer = new Accelerometer();
```

5. A new Boolean object will be used to test whether the Accelerometer works:

```
var isSupported:Boolean = Accelerometer.isSupported;
checksupport();
```

6. The following function contains the event listener that waits for the Accelerometer to be triggered:

```
function checksupport():void {
if (isSupported) {
myTextField.text = "Accelerometer feature supported";
acc1.addEventListener(AccelerometerEvent.UPDATE,
updateHandler);
} else {
myTextField.text = "howdy "; }
}
```

7. The final function posts a message to the text field to tell what direction the device has moved to:

```
function updateHandler(evt:AccelerometerEvent):void {
myTextField.text = String("at: " + evt.timestamp +
"\n" + "acceleration X: " + evt.accelerationX + "\n"
+ "acceleration Y: " + evt.accelerationY + "\n" +
"acceleration Z: " + evt.accelerationZ);
}
```

8. The final step is to package your code into an Android app and test it on your phone.

The Accelerometer gives you new ways for your customers to interface with your applications.

An added feature: the Accelerometer works great on Android devices, but the same code can be used for Adobe AIR apps running iOS, WebOS, and RIM's BlackBerry phones. Yes, that's right. Develop one app and have it deployed to multiple mobile devices. How cool is that?

Knowing Where You Are Using Geolocation

Location awareness is key to mobile devices. In this chapter you will use ActionScript to communicate with the Android's Geolocation services to determine where you are located.

Geolocation works by using satellite GPS coordinates to pinpoint your location within 4 feet of your current position. You can see how this can be useful for solutions where you need to know where you are in relationship to other coordinates.

Adobe's AIR gives you access to GPS data through the Geolocation class. Common properties you can read include:

- Latitude
- Longtitude
- Altitude
- Horizontal Accuracy

In addition to these commonly accessed properties you can also test the speed at which the phone is moving by measuring distance moved over a specific period of time.

The following example will simply post your location to your phone. What you can do with this, however, is take the data and apply it to location data. For instance, you might be writing an app where you want to see how far you are from the nearest campground.

1. Go ahead and create a new Flash movie. Name the new movie Geo.fla.

2. In the Properties panel choose the AIR Android Settings button. Select the Permissions tab.

3. From the Permissions screen, choose the hardware permission ACCESS_FINE_ LOCATION to access the phones GPS hardware.

4. On the Stage, create a new text field and label it "myTxt."

5. Select frame 1 on the Timeline, and open the Actions Panel.

6. The first step in your code is to import the frameworks you need for this example to work. In this case, the two frameworks are Geolocation and GeolocationEvent.

```
import flash.events.GeolocationEvent;
import flash.sensors.Geolocation;
```

7. The next step is to declare a new Geolocation variable. In this instance, you are
    going to name the new variable "myGeo."

```
var myGeo:Geolocation;
```

8. The following IF/ELSE statement is looking to see if Geolocation is supported.
    If Geolocation is supported, Flash triggers the myGeolocationUpdateHandler to
    access GPS information on your current location.

```
if (Geolocation.isSupported)

{

myGeo = new Geolocation();

myGeo.setRequestedUpdateInterval(100);

myGeo.addEventListener(GeolocationEvent.UPDATE,
myGeolocationUpdateHandler);

}

else

{

myTxt.text = "No geolocation support.";

}
```

9. The following function extracts data from the GPS hardware and posts the results
    to the text field on the Stage.

```
function myGeolocationUpdateHandler(event:Geolocation
Event):void

{

myTxt.text = "Geolocation is supported!" + "\n";

myTxt.appendText("latitude:" + event.latitude.
toString() + "°\n");

myTxt.appendText("longitude:" + event.longitude.
toString() + "°\n");

myTxt.appendText("Altitude:" + event.altitude.
toString() + " m\n");

myTxt.appendText("horizontal accuracy:" + event.
horizontalAccuracy.toString() + " m");

}
```

10.   Save your file, and then publish to your Android phone. You will notice a slight pause after your app loads as it collects the GPS coordinates.

As you can see, this is a very simple example of using GPS. Here are the basics from which you can now build solutions out.

Loading RSS Data into Flash

The challenge with connecting to RSS readers is the number of different RSS technologies you have out in the wild (ATOM, RSS 1, and RSS 2). This is where your knowledge of Flash can really come into play.

ActionScript 3 is not a new technology. It has been around for many years. To this end, you have a large collection of open-source libraries you can use make it much easier to create your ActionScript. We are going to do just this for one RSS reader.

The open-source code is called as3syndicationlib and is hosted at Google's Code Repository (http://code.google.com/p/as3syndicationlib/). This may sound alarming, but the latest update was in 2006. Yes, that seems like eons ago, but ActionScript 3 is at a point where it is mature. All you have to do is look in the right places.

Go to the Downloads page, download the code, and place it in the folder in which you will be creating the RSS feed. Now, open Flash CC, and create a new AIR for Android application.

1.   Save your new file. Open the AIR Android Settings and select the Permissions tab. Select the INTERNET permission. This will allow the app to load the external RSS feed.

2.   On the Stage, draw a text field, and add the ID "rssContent."

3.   Draw a new image on the Stage. Convert the image to a symbol. Give the new symbol an ID "rssButton."

4.   Open the Actions panel. The first step is to identify which frameworks you are going to need in this project:

```
import com.adobe.utils.XMLUtil;
import com.adobe.xml.syndication.rss.Item20;
import com.adobe.xml.syndication.rss.RSS20;
import flash.events.Event;
import flash.events.IOErrorEvent;
import flash.events.SecurityErrorEvent;
import flash.net.URLLoader;
import flash.net.URLRequest;
import flash.net.URLRequestMethod;
```

5.   The first ActionScript function will define the RSS feed you want to load.

```
var loader:URLLoader;

static const RSS_URL:String = "http://i3dot0.
blogspot.com/feeds/posts/default";

function onLoadPress():void

{

rssLoader = new URLLoader();

var rssRequest:URLRequest = new URLRequest(RSS_URL);

rssRequest.method = URLRequestMethod.GET;

rssLoader.addEventListener(Event.COMPLETE,
onDataLoad);

rssLoader.addEventListener(IOErrorEvent.IO_ERROR,
onIOError);

rssLoader.addEventListener(SecurityErrorEvent.
SECURITY_ERROR, onSecurityError);

rssLoader.load(request);

}
```

6.   The following action is called when the RSS data is fully loaded:

```
function onDataLoad(e:Event):void

{

var rawRSS:String = URLLoader(e.target).data;

parseRSS(rawRSS);

}
```

7.   Now you can parse your data from the loaded RSS feed. The following will post
     the title from the RSS feeds into the text field.

```
function parseRSS(data:String):void

{

if(!XMLUtil.isValidXML(data))

{

writeOutput("Feed does not contain valid XML.");

return;

}
```

```
var rss:RSS20 = new RSS20();

rss.parse(data);

var items:Array = rss.items;

for each(var item:Item20 in items)

{

writeOutput(item.title);

}

}
```

8. The following function will post the data to text field on the Stage:

```
function writeOutput(data:String):void

{

rssContent.text + = data + "\n";

}
```

9. The following functions will output any errors you receive to the text field:

```
function onIOError(e:IOErrorEvent):void

{

writeOutput("IOError : " + e.text);

}

function onSecurityError(e:SecurityErrorEvent):void

{

writeOutput("SecurityError : " + e.text);

}
```

10. The final step is to add a listener that will trigger the RSS feed to load.

```
rssButton.addEventListener(MouseEvent.CLICK,
buttonClick);

function buttonClick (e:MouseEvent):void{

onLoadPress();

}
```

11. Save your file, and test on your Android device.

You can now load external data, in the form of RSS, from a different website. This is a really big deal for your development, as it demonstrates that you can integrate data from different sources.

Adding Permissions to Your Apps

The processes for developing iOS and Android apps can be slightly different. The main difference is that you can currently do more with Android apps hardware than with iOS. Will this change over time? I am certain it will, but this is where we are for now.

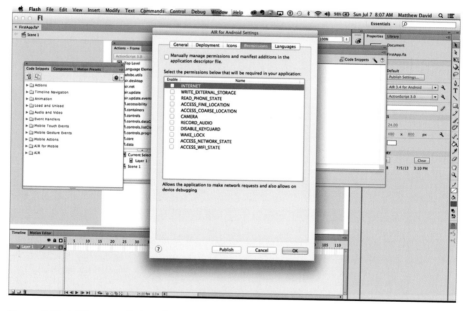

Figure 5.13: Adding permissions for Android apps.

Many of the Android-specific features listed in the following examples require you to activate specific permissions in your code. Fortunately, this is easy to do. Select AIR ANDROID settings on the Properties panel. The Application and Installer window opens. Choose the fourth tab along the top, labeled Permissions. You will see a whole list of permissions you must select if you are going to use the hardware on the device.

The following is a list of permissions you can select:

- INTERNET
- WRITE_EXTERNAL_STORAGE
- READ_PHONE_STATE
- ACCESS_FINE_LOCATION
- CAMERA

- RECORD_AUDIO

- DISABLE_KEYBOARD

- WAKE_LOCK

- ACCESS_NETWORK_STATE

- ACCESS_WIFI_STATE

Each of these hardware-specific permissions controls a different element of your Android phone. Some are obvious, such as RECORD_AUDIO to control the microphone. Some are less obvious, such as ACCESS_FINE_LOCATION to activate the GPS settings.

You can select which permission you need for your application. You do, however, have another way to modify which permissions you can use.

Each application you develop will create an XML configuration file. The role of the file is to define launch settings, code location, and other features. One element is called the MANIFEST. The Android manifest lists in XML the hardware permissions you can use. The following demonstrates how to add permission to use the CAMERA:

```
<android>
<manifestAdditions>
<manifest>
<data><![CDATA[<uses-permission android:name = "android.
permission.CAMERA"/>]]></data>
</manifest>
</manifestAdditions>
</android>
```

It is certainly easier to use the UI interface in Flash to state which permissions you want to use, but knowing that you can manually access and modify the permissions with your favorite notepad does have its benefits. For instance, you can add reference to a new hardware permission that may not have made its way to Flash CC UI. An example of this is support for VIBRATE.

Loading Web Pages into the StageWebView

It is important to remember that AIR for Android is not just a crippled version of the Adobe Integrated Runtime but almost the complete version of AIR. A key part of AIR on the desktop is the ability to launch Web ports and to pull complete Web pages into your Flash world. Well, AIR on Android will do the same, and it is crazy easy to implement.

There are two key elements you need to keep in mind when using StageWebView:

- Ensure you have set your App permissions correctly.

- Use the StageWebView object.

The first step is to create a new AIR for Android application and open the AIR for Android settings in the Properties panel.

Choose the fourth tab, labeled permissions. The WebView object will load an external Web page. For this to occur you must select the INTERNET permission option on the Permissions tab. If you do not do this, you will not be able to load a Web page.

Alternatively, you can manually update the XML manifest document. The following will allow the INTERNET permission to work:

```
<android>

<manifestAdditions>

<manifest>

<data><![CDATA[<uses-permission android:name = "android.
permission.INTERNET"/>]]></data>

</manifest>

</manifestAdditions>

</android>
```

The next step is to add the ActionScript that will load a StageWebView. Let's do that right now.

1. Select frame one in the timeline. Open the Actions panel.

2. Let's start by creating a new StageWebView object and naming it "webView."

    ```
    var webView:StageWebView = new StageWebView();
    ```

3. The following code states that the new StageWebView object will reside on the current Stage:

    ```
    webView.stage = this.stage;
    ```

4. You can define the position and size of the Web view you load. This is called the "viewport." The following ActionScript creates a new rectangular viewport and places it in the top left corner with a width of 470 px and a height of 300 px.

    ```
    webView.viewPort = new Rectangle(0,0,470,300);
    ```

5. Finally, you need to state the Web page you want to load using the loadURL property.

    ```
    webView.loadURL("www.google.com");
    ```

6. Save your file, and then load it onto your Android phone. You will see that the AIR app opens and reveals Google's home page.

There are some caveats when working with StageWebView. The first is that you cannot communicate between Flash and the Web page. Second, the Web browser used to load the page is not the default Android browser but a branched version of WebKit. This means you do not have the JavaScript acceleration the V8-powered Android browser has. Finally, the StageWebView does take over the area of the Stage it is loaded onto. This means you lose space you could otherwise use for application development.

There are, however, lots of benefits to running StageWebView. It is, after all, an easy way to integrate Flash and HTML. In addition, you can load two or more viewports to a screen. For instance, you can add the following code to include a second viewport:

```
var webViewTwo:StageWebView = new StageWebView();

webViewTwo.stage = this.stage;

webViewTwo.viewPort = new Rectangle(0,305,470,300);

webViewTwo.loadURL("www.focalpress.com");
```

You are also not restricted to loading Web pages on external sites. You can load pages locally to the application. For this, however, you must remember to include the local HTML as part of your application when you build the app. This is done through the AIR Android settings tab. Choose the folder with the included HTML you want to include in the final package.

Controlling the Use of the Microphone

The microphone is arguably the most used part of a phone. You need it for every call you make. You can also use it to record audio. The following section demonstrates how you can leverage the microphone in your solutions.

1. Start by creating a new Flash AIR for Android solution.

2. Open the AIR Android settings, and select Permissions. Check the AUDIO permission.

3. On the Stage, select frame one from the timeline, and open the Actions panel. Add the following to create a 4-second delay in your project:

```
const DELAY_LENGTH:int = 4000;
```

4. The next step is to create the new microphone object. Here you will see that it has been abbreviated to "mic."

```
var mic:Microphone = Microphone.getMicrophone();
```

5.   Two properties for the microphone include gain (loudness) and rate (sound quality).
      The following sets the gain and rate for the new mic object.

```
mic.gain = 100;
mic.rate = 44;
```

6.   The following sets the microphone to stop working (silence level of 0) after the
      DELAY_LENGTH period has passed. In this instance, DELAY_LENGTH is 4,000
      milliseconds, or four seconds.

```
mic.setSilenceLevel(0, DELAY _ LENGTH);
```

7.   The following line triggers the microphone object listener.

```
mic.addEventListener(SampleDataEvent.SAMPLE _ DATA,
micSampleDataHandler);
```

8.   The following timer object uses the same time-based technology in ActionScript
      covered in the previous section, demonstrating that you do not need to relearn
      Flash to build Android apps.

```
var timer:Timer = new Timer(DELAY _ LENGTH);
timer.addEventListener(TimerEvent.TIMER, timerHandler);
timer.start();
var soundBytes:ByteArray = new ByteArray();
```

9.   The following function captures the sound to the phone's memory.

```
function micSampleDataHandler(event:SampleDataEvent):void
{
while (event.data.bytesAvailable)
{
var sample:Number = event.data.readFloat();
soundBytes.writeFloat(sample);
}
}
```

10.   The following uses the mic object to record a new sound file.

```
function timerHandler(event:TimerEvent):void

{

mic.removeEventListener(SampleDataEvent.SAMPLE_DATA,
micSampleDataHandler);

timer.stop();

soundBytes.position = 0;

var sound:Sound = new Sound();

sound.addEventListener(SampleDataEvent.SAMPLE_DATA,
playbackSampleHandler);

sound.play();

}
```

11.   The following function will play back the four seconds of recorded audio.

```
function playbackSampleHandler(event:SampleDataEvent):
void

{

for (var i:int = 0; i < 8192 && soundBytes.
bytesAvailable > 0; i++)

{

var sample:Number = soundBytes.readFloat();

event.data.writeFloat(sample);

event.data.writeFloat(sample);

}

}
```

12.   Save your work. Compile and load the APK file onto your Android phone. Select the app, and talk into your microphone. After four seconds, the audio will stop and will play back to you.

You can easily extend this example. For instance, you can add a button that allows you to click and record the audio; audio files can be saved to the physical hard drive on the Android phone; you can even use the many ActionScript libraries that modify sound files to create a sound modulator. In other words, you can do a lot.

Controlling the Camera

For me, one of the coolest features you can access on your phone is the camera. The goal of this section is to demonstrate how you can access the camera on your Android phone. As with the microphone example earlier, you can add additional ActionScript that will allow you to save your video for playback later or even add color correction controls.

But enough of that; let's jump right into the project.

1. The first thing is to create a new Android Flash project and then associate the correct hardware permissions. You should be comfortable doing this by now. The hardware that needs permission is the "CAMERA."

2. Instead of adding the code into the Timeline, let's go ahead and create a simple class for the AIR solution. Name a new class in the Properties panel takeVideoTest, and select Flash as the code environment.

3. After the opening package, add the following references to different frameworks.

```
package
{
import flash.display.Sprite;
import flash.media.Camera;
import flash.media.Video;
import flash.text.TextField;
import flash.text.TextFieldAutoSize;
import flash.text.TextFormat;
import flash.utils.Timer;
import flash.events.TimerEvent;
import flash.events.StatusEvent;
import flash.events.MouseEvent;
import flash.system.SecurityPanel;
import flash.system.Security;
```

4. Next, let's declare the variables you will be using:

```
public class takeVideoTest extends Sprite
{
private var myTxt:TextField;
private var headerTxt:TextField;
private var cam:Camera;
```

```
private var t:Timer = new Timer(1000);
public function takeVideoTest()
```

5. The variables listed here control two different text fields, the camera and a timer control.

6. The next block defines the size, position, and other properties of the myTxt field.

```
{
myTxt = new TextField();
myTxt.x = 10;
myTxt.y = 10;
myTxt.background = true;
myTxt.selectable = false;
myTxt.autoSize = TextFieldAutoSize.LEFT;
```

7. The following ActionScript defines the properties of the headerTxt field.

```
headerTxt = new TextField();
headerTxt.x = 120;
headerTxt.y = 220;
headerTxt.autoSize = TextFieldAutoSize.LEFT;
```

8. The following is a style document that formats the visual presentation of the text fields:

```
var format:TextFormat = new TextFormat();
format.font = "_Sans";
format.color = 0xFF0000;
format.size = 24;
format.bold = true;
headerTxt.defaultTextFormat = format;
addChild(headerTxt);
```

9.   The following IF/ELSE statement is looking to see if the camera is installed. Remember, while it is common for Android phones to have a camera, it is not mandatory. The first part of the IF statement will throw a message if there is no camera installed.

```
cam = Camera.getCamera();
if (! cam)
{
myTxt.text = "No camera is installed.";
}
```

10.   If the camera is installed and is working, the following message will be sent to the myTxt field informing the user that the camera is connecting.

```
else
{
myTxt.text = "Connecting";
connectCamera();
}
addChild(myTxt);
t.addEventListener(TimerEvent.TIMER, timerHandler);
}
private function clickHandler(e:MouseEvent):void
private function statusHandler(event:StatusEvent):void
{
if (event.code = = "Camera.Unmuted")
{
connectCamera();
cam.removeEventListener(StatusEvent.STATUS,
statusHandler);
}
}
```

11.   The following function controls the size and position of the video playback.

```
private function connectCamera():void

{

var vid:Video = new Video(cam.width,cam.height);

vid.x = 10;

vid.y = 10;

vid.width = 120;

vid.height = 120;

vid.attachCamera(cam);

addChild(vid);

t.start();

}
```

12. Finally, the following function will send data about the video cameras performance to the myTxt screen. Video frames per second playback will vary depending on your hardware.

```
private function timerHandler(event:TimerEvent):void

{

myTxt.y = cam.height + 20;

myTxt.text = "";

myTxt.appendText("bandwidth: " + cam.bandwidth +
"\n");

myTxt.appendText("currentFPS: " + Math.round(cam.
currentFPS) + "\n");

myTxt.appendText("fps: " + cam.fps + "\n");

myTxt.appendText("keyFrameInterval: " + cam.
keyFrameInterval + "\n");

headerTxt.text = "Video Camera Test";

}

}

}
```

13. The final step is to save your work and then test it on your Android phone.

Video support in AIR is going to be a big deal. Through access to the camera you can add augmented reality to your AIR apps, video editing, and video conference similar to Apple's FaceTime.

Preparing Your Application for Deployment into the World

You've done it. You have your app ready to go into the wild and make some money. Adobe's Flash with ActionScript give you alternative ways to build your applications for distribution to Apple's iTunes App Store or to Google Play.

Developing Apps with Zero-Code Tools

Tools are emerging that allow anyone, no matter how advanced his development skills, to create native applications.

The focus of this chapter is to illustrate that the world of mobile application development is for everyone. "No Code Needed" is the tag line you can hang on the bottom of each page. There are a growing number of companies that are developing code-free app tools through plug-and-play interfaces. The tools are easy to use.

There are three tools we will zero in on when it comes to creating apps:

- TheAppBuilder
- GameSalad
- Appery.io

All three tools target different types of apps you can create. TheAppBuilder is for casual content apps, GameSalad gives you the tools to build games, and Appery.io is a tool for building enterprise solutions.

In addition, I have added a section on iBook Author. While it is not technically an app, the line between interactive eBook and app is getting very blurred. iBook Author gives you the tools to create highly immersive apps, and when you add the design guts of iAd Producer, you can create extraordinary solutions—all with zero code.

Figure 6.1: Shadow Run, by Utopian Games, was written without a single line of Objective-C code for the iPhone.

OK, to be fair, the solutions you create here will not be of the same quality as Angry Birds, but you will be surprised at how good your solutions will be. If nothing else, building your first app with a zero-code tool gives you tremendous experience in understanding how apps work and how your business will work using an app.

TheAppBuilder.com

Not all startups need to be from Silicon Valley. JamPot, a company located in Belfast, Northern Ireland, has been making quite a splash over the past couple of years by creating innovative technology that mere mortals can use.

Figure 6.2: JamPot is an exciting startup out of Belfast, Northern Ireland.

Figure 6.3: JamPot's TheAppBuilder is a Web-powered app authoring tool.

The latest tool from JamPot is TheAppBuilder, a tool that you can use to create mobile apps for iOS, Android, HTML5, and Windows Phone without writing a line of code. The goal of TheAppBuilder is to make it as easy as possible to create an app. And, when you use the tool, you may well agree with them.

RAPID APP CREATION

App creation using TheAppBuilder requires only a Web browser. That's it. Let's step through the process you need to complete to create an app with TheAppBuilder.

1. Start by going to TheAppBuilder.com.

2. Select the "Get Started, It's Free" button on the home page.

3. A new window will open that offers a set of predefined templates you can use to kick-start you app development process.

4. Let's create an app for your business. Select the "Business" tab.

5. Choose the "Restaurant" template. Select the "Next" button.

6. On the next screen, enter your contact information. That's it; you have a template app created.

7. You will now be sent to the edit screen, where you can edit content for your app in real time.

Amazingly, you can edit almost everything on the app (the splash screen, the app name, the icons, all of the content), and when you are done, you can keep adding content after you app has been published.

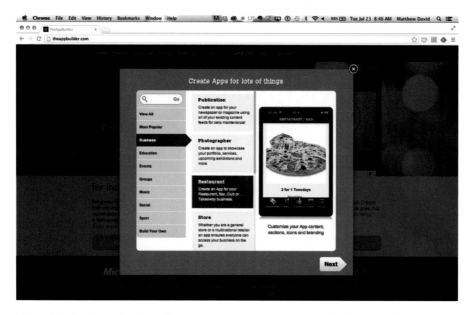

Figure 6.4: JamPot makes it really easy to create your first app with TheAppBuilder.

NEW CONTENT

TheAppBuilder is constantly adding new features, such as maps, forms, and lists. Each content piece is optimized for each platform you publish to, such as Android and iPhone. This means that the Human Interface Guidelines (HIG) are supported for each platform.

DEVELOPMENT FOR ALL PLATFORMS

Apps you create with TheAppBuilder will work on all of the leading mobile platforms (iPhone, Android, and Windows phone).

A great way for you to test out your app is to view it on your phone with the mobile-optimized website. Yes, you get a mobile website with your app, too! Your Web address is http://myapp.is/*nameofyourapp* and is free with the service.

EXTENDING THEAPPBUILDER

Apps created with TheAppBuilder can be extended with custom HTML5 connected to the app. The good news is that mobile browsers have tremendous support for HTML5 that gives you the freedom to include almost any type of content in your app.

Figure 6.5: Instant publishing to all leading app stores is possible with TheAppBuilder.

BENEFITS AND RESTRICTIONS OF THEAPPBUILDER

TheAppBuilder is a great place to start with your first app. It is cheap (free to try and $30 a month to publish to leading app stores) and is going to be a great solution for small companies, organizations, nonprofits, and individuals who want to dabble. What TheAppBuilder is not is a tool for enterprises and larger organizations. The tool is simply too restrictive for more complex solutions.

The lads from Belfast are working hard, and I am sure they will address the restrictions over time. For now, you have a great tool for proving a concept or starting your first mobile app for your business, nonprofit, or local organization.

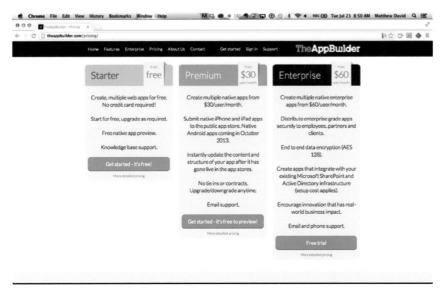

Figure 6.6: TheAppBuilder, which lets you update your app, is free to try and only $30 a month thereafter.

GameSalad

C'mon, you know you have some games on your phone. Is it Angry Birds, Temple Run, or Doodle Jump? Each of these games has generated millions in revenue for its publisher. Indeed, Angry Birds is the world's most popular game of all time, with more than 1 billion downloads. That's more than every game of Nintendo's Mario put together.

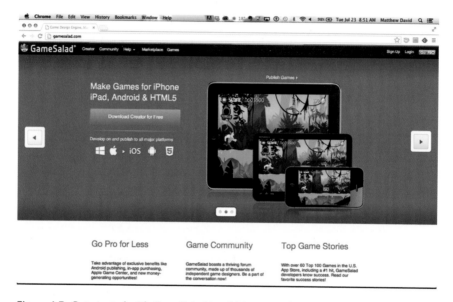

Figure 6.7: Get started with GameSalad by visiting its website.

GameSalad is a cool company that has developed a set of tools you can use to create your own games. Again, not a snitch of code is needed. Just bring your imagination.

GAMES, GAMES, GAMES

There is only one focus to GameSalad: games. The company wants you to focus only on game development. And I think this is a great idea. There are too many "we can do everything" solutions on the market, so it is refreshing to come across one company that focuses on one area.

Currently the games you can develop with GameSalad are all 2-D games, but you will be amazed at what you can create. Games such as Angry Birds, Where's My Water, and Plants vs. Zombies can all be duplicated with GameSalad. OK, this does not mean that you duplicate the games, but it does mean that you can use the game mechanics in your own games.

All of your development is through the GameSalad Creator, a free download on GameSalad. com. Those of you who remember the good ol' days of Macromedia Director will recognize the breakdown of characters and scenes in GameSalad. The idea is that you can add "characters" that you control to scenes where the character interacts.

Figure 6.8 GameSalad comes packaged with templates to help you get started.

It is best to start off your first game as something simple. Use the great online tutorials (http://cookbook.gamesalad.com/tutorials) to step you through the process of using the GameSalad tools. You will learn everything you need to build a game.

What you will like about GameSalad is that it gives you easy access to features such as adding achievements, in-app purchases, and advertising. There simply is no other

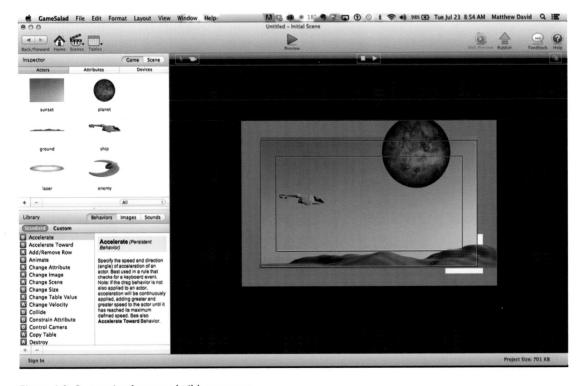

Figure 6.9: Creator is where you build your game.

tool on the market that gives you this level of control without requiring that you be a professional developer. Some of the advanced features are available only if you have a professional license for GameSalad Creator. The license is only $299 a year—an incredible deal.

LEVERAGING TEMPLATES AND AN ACTIVE COMMUNITY

Almost every aspect of the scene and characters is controlled with settings. There are a lot of settings! My biggest criticism of GameSalad is the complex settings. Fortunately, you can buy templates that can kick-start your work.

Templates, artwork, and game music can be purchased through the GameSalad market. Some of the best templates to start using are those developed by GameSalad, and they are amazing.

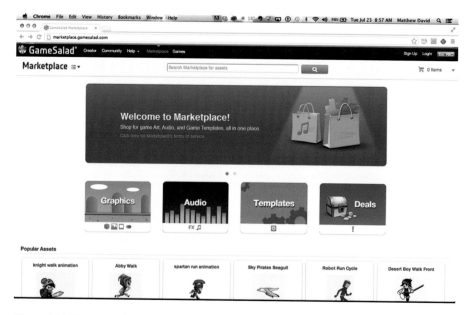

Figure 6.10: There are dozens of prebuilt templates you can download to jump-start your game creation.

You will find that the forums are a great place to hang out. There are two levels of forums: free and Pro. The forums have a lot of great helpers, a reflection of the game development community as a whole.

PUBLISHING YOUR GAMES

After you have created your game, you will want to test it. GameSalad supports a huge number of platforms for you to publish to, including iPhone, iPad, Android, and Windows Phone. In addition to the mobile platforms, you also have support for HTML5, native Apple OS X, and Windows 8.

Figure 6.11: The forums have a passionate and active community that is always willing to help.

What is really cool is that the games you create are optimized for each platform. Games are not simply stretched to fill a screen. With all of these screens and operating systems, you do need to complete some testing. GameSalad has you covered here, too. Built into GameSalad are all of the emulators you need for every platform. Testing on Android, iOS, and Windows has never been easier.

GameSalad publishes all of the games you create in the cloud. This eliminates your having to have all the software needed to compile completed games on your computer. The end result is that you have a game that you can take and publish to any of the leading app stores.

THE GROWING NUMBER OF APP STORES

The number of places where you can find apps is growing dramatically. One new app store is created every week. Stores are typically tied to platforms or devices. There are stores for phones, tablets, TVs, Blue-Ray players, game consoles, in-car systems—the list goes on and on. To reach all of your potential customers, you may want to consider submitting your apps to more than just the standard app stores.

BENEFITS AND RESTRICTIONS OF GAMESALAD

Keep your eye on GameSalad. It is advancing very rapidly. Already it has support for Tizen, Samsung's latest mobile operating system. If you want a tool that will publish games to all leading platforms, then this is it.

Figure 6.12: GameSalad boasts a large and growing collection of games.

Currently, however, there is no support for 3D in your games. Mobile devices are becoming more and more powerful. I recently saw a demo of a phone running a console-quality 3D game. Amazing! At some point, GameSalad will need to add 3D support to allow hobbyists to migrate from casual user to full-time game creator.

The potential is there for you to build a good game with GameSalad. What I want to see is the potential to build a great game!

Appery.io

The majority of this book focuses on consumer apps that can be installed from popular app stores such as Apple's iTunes App Store, Google Play, or Amazon's App Store. The next section takes looks at how you can create apps for your enterprise. Enterprise apps are one of the fastest-growing sectors in mobility.

Enterprise apps give you the ability to release apps that are tied to proprietary tools and systems. Many companies are starting to develop their own enterprise apps to allow employees and partners to access corporate systems from a phone or tablet.

Currently very few companies are producing tools that allow you to build enterprise apps. One of the leaders is Appery.io, which operates a 100 percent cloud solution.

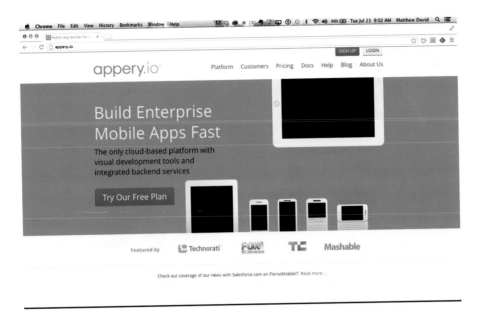

Figure 6.13: Appery.io is an enterprise app creation tool.

ENTERPRISE APP STORES

Any company can create its own app store. For Android you do not have to change how you develop your apps. They will just work. For iOS you will need to add an Enterprise Certificate (which must be renewed every 12 months). The advantages of an enterprise app store is that you control which apps are published and which are not.

BUILDING APPS IN THE CLOUD

Appery.io is powered 100 percent through the cloud. Your account is managed in the cloud, your apps are designed in the cloud, and your apps are built in the cloud. All you need is a Web browser. If you wanted to, you could build your solutions on an iPad using the Safari Web browser.

The first step is to register with the website. From there you can select what type of app you want to create.

To get you started, I encourage you to step through the RSS Reader app. This app will give you the knowledge you need to get up and running building an app with Appery.io. The link to the tutorial is http://docs.appery.io/tutorials/building-a-mobile-rss-app/.

GETTING STARTED WITH APPERY.IO

Appery.io has a great set of tutorials you can use to get up to speed very quickly with its tool. Here is the Web address: http://docs.appery.io/.

Figure 6.14: All apps are created in the cloud with PhoneGap.

A version of PhoneGap is managed by Appery.io in its cloud to let you build apps for all supported platforms without having to have all the software installed on your computer.

EXTENDING APPS WITH PLUGINS

Again, the focus of Appery.io is the enterprise. There is a growing number of plugins that can extend how you leverage Appery.io apps in your company. The plugins have a distinct enterprise focus. You will find plugins that let you connect to SalesForce, Box, and security models such as Oauth.

The price for the plugins vary. Some are free, and some require a license from the service plugin author. Check the small print before you commit.

BOOTSTRAPPING YOUR APPERY.IO APP WITH HTML5

The code base that powers Apper.io is HTML5. The build server is PhoneGap. So, knowing that PhoneGap is the wrapper for your apps and that you can use JavaScript, CSS, and HTML with PhoneGap gives you options to extend the foundation app you create with Apper.io.

BENEFITS AND RESTRICTIONS OF APPERY.IO

Appery.io has a great interface that lets you rapidly build complex applications integrated into corporate systems. The challenge you are going to run into with Appery.io is taking solutions beyond the drag-and-drop framework. Appery.io is easy to use and is great for proofs of concept. Solutions that publish HTTP REST, such as those published through SAP Netweaver Gateway or from Microsoft's SharePoint Server, are perfect candidates for optimizing with an Appery.io mobile interface. More complex solutions where you leverage hardware and native OS APIs are not good candidates for Appery.io.

Creating Interactive Books

This will come as a surprise to you all: I am a writer. OK, so that is cool. What is really interesting today is the massive range of options a writer has. Publishing a book that people can buy at Barnes and Noble is not the only option.

In this section I am going to jump into the concept of the digital book. The reason for this is that the digital book is much more sophisticated today than just a reprint of your published book in e-ink format. In many ways, a fully interactive book can contain the same level of complexity as any app outlined in this chapter so far. The leader for interactive eBooks is Apple, with its iBooks app for the iPad and Mac OS X.

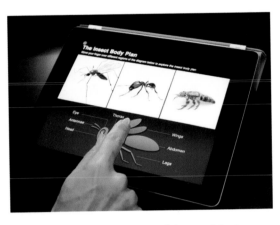

Figure 6.15: Apple is bringing rich interactivity to digital books.

The bottom line is that today's interactive books give you the option to deliver highly engaging solutions without the need for a developer. You are the master of interactive control.

CREATING INTERACTIVE BOOKS FOR ALL PLATFORMS

Apple's iBook Author has a specific target: Apple products. To reach other platforms such as Android, you will need to use other products. If you are an Adobe Creative Cloud subscriber, then you already have a choice: Adobe's InDesign. InDesign is a page-layout product similar in concept to Quark. The latest release of InDesign now lets you add interactive elements to your designs. Check out Martha Stewart magazines (available on any device you have in reach), and you will see what you can do with InDesign. It is very powerful. The downside is the learning curve.

DIGITAL BOOKS AS ALTERNATIVES TO APPS

Five years ago the digital book market hardly existed. Then Amazon released the Kindle. Wow, that was a good move. People love to carry their libraries around in their pockets. The long battery life of the first Kindle is awesome. A friend of mine goes to a Caribbean island for five weeks every other year with his wife and his collection of books. The last couple of trips, he just took a Kindle loaded with all the books he wants to read. He did not bring a charger with him. The low power of the Kindle was enough to let him read four to eight hours a day for five weeks. This is why the Kindle is so popular.

Figure 6.16: Adobe's InDesign has the tools for you to create interactive books for a broad range of devices and bookstores.

I have kids, so the best I can manage is a tent in the backyard for one night. My friend going to the Caribbean is a much more romantic story.

The Kindle is very much the first version of a device for digital books. There are many standards for digital publishing: MOBI (used on Kindle), PDF (standard defined by Adobe), and EPUB (supported by Google, Apple, Microsoft, Sony, Samsung, and many other companies). A comprehensive list of devices and supported technologies is maintained at this Wikipedia entry: http://en.wikipedia.org/wiki/Comparison_of_e-book_readers.

The technologies are evolving fast. Many of the latest devices now support rich media and interaction with support for HTML5 right in your books. This gives you a foundation for building great solutions.

iBOOK AUTHOR

The tool you use to create a book for sale through Apple's iBooks app store is iBook Author, which uses the ePub 3.0 standard that allows you to add amazing interactivity. But the books you create will run only on iPads and Mac OS X. So how big is your audience? Today, there are more than 150 million devices (iPad, iPad Mini, and Macs) that can view the books you create. Big enough that there are more than 10,000 books published just for the iPad and Mac. That's a lot in anyone's book.

WHERE iBOOKS AUTHOR DOES NOT WORK

Creating interactive books with iBooks Author is fun, but there is a big caveat: your books will not work on the iPhone. All of your content will run great on iPad, iPad Mini, and any Mac running OS X Mavericks, but not on the iPhone or iPod Touch.

Figure 6.17: Apple's iBook Author is used to publish interactive content.

iBook Author runs only on Macs, but it is free through the Mac App Store. A huge advantage to buying software through Mac App Store is that the software always updates itself to the latest release. There is no need for you to keep track of whether you have the latest features.

To get you started on your next great book, you need only open iBook Author. The first screen is a set of templates that lets you choose the style of book you would like to write. Templates come either as Landscape with Portrait or Portrait Only. The Landscape templates are great for writing books that will be viewed on Macs and full-sized iPads. The iPad Mini is great for Portrait solutions.

Figure 6.18: Apple offers a number of iBook Author templates.

BUYING NEW TEMPLATES

There is a growing market of templates you can buy to extend the base set of templates available through iBook Author. The best way to find the new templates endorsed by Apple is to search for iBook Author Templates in the Mac App Store. There are currently a dozen different companies developing templates you can download and use with iBook Author.

You will see that there are three main sections to iBook:

- The far left displays an outline of your book.
- The main section of the screen shows the content and presentation of your content.
- Across the top are the additional widgets and controls for your content.

The structure of your book is that of chapters and publishing content. In many ways, Apple forces you to think like a professional publisher.

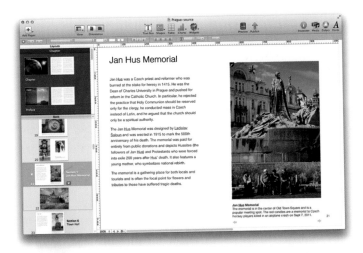

Figure 6.19: The design environment in iBook Author.

As you might expect, written content can be added directly to the page. Just start writing. You will see familiar tools to format your written content. Additionally, you can import content written with Apple's Pages. Pages can be used to edit grammar and spelling.

The fun comes when you want to add interactive content to your book. You will find all sorts of great options for interactive content in iBook Author, including:

- T-charts and lists
- Image gallery
- Video and audio clips
- Review questions
- Keynote presentations
- Interactive images
- 3D
- Scrolling sidebar
- Pop-over text box

Each of these tools is easy to learn and master. Just drag and drop the content onto the screen.

Figure 6.20: Widgets add inter-activity to the books you create.

HTML5 AND iAD PRODUCER

There is an additional tool you can use to create rich interactivity in your books: HTML5. The HTML5 widget gives you the option to embed any supported HTML5 elements into your book. It becomes easy to embed live elements such as Twitter feeds directly into your book. Leveraging HTML5 in your iBook allows your book to consume live content and converts your book from a static object captured in time to one that is living and growing with new content pulled from the Web.

Complex HTML5 can be difficult to do without getting your hands dirty with programming. Urgh! This chapter is not about that. Apple's iAd Producer is a tool you can use that lets you build complex animation in HTML without learning a single line of code.

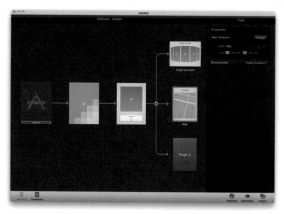

Figure 6.21: The iAd Producer Story Book layout.

The core focus for iAd Producer is to create interactive advertising for iPhones and iPads. The ads are created with HTML, images, CSS, and JavaScript. What is cool about iAd Producer is that it is one of the best-unknown animation tools for HTML5 you can get. And it is free!

To get started, you will need to download iAd Producer. As you might expect, iAd Producer runs only on Macs. You can find out how to download and install iAd Producer here: https://developer.apple.com/iad/iadproducer/.

Apple's "Story Book" process is used to create interactive elements in iAd Producer. Cards split up each screen you work on. Each screen can then have complex animation and interaction added. No code is needed. All of the scripting and management of images and CSS are completed by iAd Producer.

When you have completed your work, you have only to select File ◊ Export and save the "widget" to your Mac desktop. Then, drag the widget into your iBook. That's it—complex HTML5 interaction is now running in your book.

USING iAD PRODUCER TO GENERATE WIDGETS FOR iBOOK AUTHOR

Check out this great "How To" document put together by Apple that steps you through what you need to do to create widgets for iBook Author with iAd Producer: http://support.apple.com/kb/HT5743.

TOOLS TO CONVERT HTML5 TO WIDGETS FOR iBOOKS

A great HTML5 animation tool that I have been using for a while is a product called Hype. It is available in the Mac App Store. A recent new feature is the ability to export any animation you create in Hype as a packaged widget for iBook Author.

PUBLISHING YOUR iBOOK

There are two main ways to share a book you create with iBook Author. The first is to send it via e-mail. You can send the file in iBook Author format (for viewing through the iBook app on your Mac or iPad), PDF, or plain text.

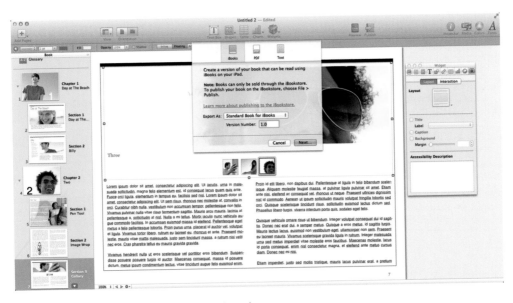

Figure 6.22: You can share books from iBook Author.

You can also publish directly to Apple's iBook Store from your Mac. You will need a developers account (register at developer.apple.com). Free books can be published with no additional work. If you want to charge money for your book (and who doesn't?), you will need an ISBN number for your book. The ISBN number is the number you see on the back of every book you have bought. To buy an ISBN number, you go to ISBN.org.

Selecting File → Publish will open the wizard that takes you through the process of packaging your book and sending it to the Apple Books store for review and publication. You will need some assets related to your published book, such as sample pages, name, description, and price.

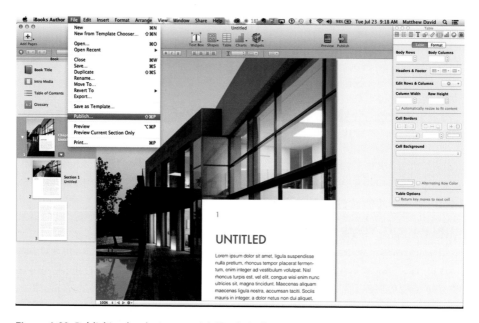

Figure 6.23: Publishing books is easy with iBook Author.

Apple has tools to measure how well your book is selling. What I find really exciting is that you can version your books, just like an App. Get version 1.0 out the door and then keep adding to it. Updates are automatically pushed out to anyone who has bought your book.

EASY DIGITAL PUBLISHING

Want to take the fuss out of publishing? The easiest way is through a free service run at LuLu.com. All you need is a Word version of your book, some artwork for the book cover, and a price you want to charge. LuLu will convert your Word file into the correct format for the leading digital bookstores (Amazon, Barnes and Noble, and Apple) and will take a small percentage of each sale. It is really easy and worth trying.

BENEFITS AND RESTRICTIONS OF INTERACTIVE BOOKS

The most significant advantage of Apple iBook Author is its ease of use. If you are comfortable using PowerPoint or Keynote, then you are 50 percent of the way to creating engaging solutions with iBook Author. The publishing process to the iBook Store is also incredibly easy if you do not want to include an ISBN number.

The fundamental restriction of using Apple's iBooks is that the target audience is just Apple users (and only those with iPads and Macs). At the time of writing, the interactive books you create with iBooks Author will not run on iPhones and will certainly not run on non-Apple devices such as Windows and Android devices.

What You Can and Cannot Do with Zero-Code Tools

There are an increasing number of options for creating complex apps that require no code for many different devices. From Web-based tools to software you install on your desktop; there seem to be no limits to the options available to you.

And this is OK for smaller projects. Remember FrontPage? The bane of all professional Web developers is Microsoft's FrontPage, a tool that made it crazy easy to create websites. The problem is that every FrontPage website looked like it had been created with FrontPage. The same challenge faces zero-code tools. It is all too easy for apps to start looking the same. As with FrontPage, "sameness" is OK for small projects, internal solutions, or places where you have little or no budget. "Sameness" is not OK for apps that have large audiences.

Be sure you fully understand your end user when you choose a zero-code app. Prototypes work great with zero code, and interactive books can be created only with designer tools (InDesign and iBook Author). Each of these tools is great in the right place.

The challenge for any mobile developer today is that you cannot learn only one set of technologies. You need a collection of tools to be successful. Zero-code tools target a range of customers that you want to service but who may have restrictive budgets.

Challenges and Successes of Native iOS Development

During the keynote address at the World Wide Developer's Conference in June 2013, Apple CEO Tim Cook announced to a sea of rabid coders that the company had paid more than $10 billion to developers just like them since 2008. Half of that had come in the past year. In that same time period, there were more than 50 billion downloads from the iOS App Store.

With figures like that, it can seem like a no-brainer to start your mobile development with the "i" platform of devices. The reality, unfortunately, can seem much trickier to independent dreamers thinking of the next mobile phenomenon.

According to a mid-2013 study released by App Promo, 67 percent of iOS developers lose money on the work they bring to market. When you factor in the hardware, software, and licensing fees, not to mention time involved, that stat can seem reasonable.

Apple's approval process gets a lot of flak in the press for being a "walled garden" where apps get rejected left and right, but the truth has been overblown a bit. Most apps get approved, even if they need a tweak here or there. Once you get approved, however, the job is not necessarily done. With nearly 1 million choices for users, differentiating your offering from the X-coder down the street takes strategic effort.

Figure 7.1: Apple's announcement of iOS 7 at the 2013 WWDC brought the most aggressive update to the platform since its inception.

HANDY NEW iOS FEATURES

If you don't have the latest additions to developer integration in iOS 7, here are some of my current favorites:

- AirDrop: Even Android users think all the Samsung commercials with phones touching each other are ridiculous. Now, iOS users can take a single piece of media and share it with everyone in the room in one action.

- Native App Improvements: Everyone I know with an iPhone talks about disliking native apps such as Weather, Mail, Calendar, iMessage, and so on. We have all read blog post after blog post of how you can replace them with cooler third-party versions. Luckily, we can put that conversation on hold (at least for a while) because of beautiful updates.

- iTunes Radio: I have a folder on my home page for all the different streaming radio services. Could this be my opportunity to put them all on hold with Apple's foray into streaming music?

- Updated Notifications: Beyond accessing my notifications from the lock screen, this is by far the aspect of iOS that needed some TLC. Users can access their day at a glance from the Notification Center in addition to the ones they missed. The three-tab configuration will come in handy.

- More Siri: Not only has the interface been improved with graphic elements of voice searches, but you can change your settings from Siri. I'm not sure I care about whether a male or female is doing the work for me, but I do think that the less "digitized" versions of Siri will help the human-interface side of the product.

Now That We Have That Out of the Way . . .

iOS users bring some advantages to the table; iPhone users are slightly younger and more affluent than Android users. According to a December 2012 study by comScore, nearly 20 percent of iOS users are between 18 and 24 years of age (compared to 16 percent for Android users). Apple also holds nearly a 2:1 advantage in household income, with 41 percent of users falling into the $100,000+ income segment (against 24 percent for Google's Android). Of course, this shouldn't be a surprise when you consider that Apple devices command a higher price point than those of other makers.

iPhone users also are more likely to pay for content than Android users. In the same comScore study, 23 percent of users said they would pay for content on their mobile devices, whereas only 17 percent of Android users said the same. Apple also boasts of an integrated payment system already installed in its operating system.

With all of the fragmentation of devices and forked versions of Android, handset makers make their users jump through a few too many hoops to capture payment data and therefore don't facilitate the actual buying process. Apple has a significant advantage with its singular Apple ID. With a single e-mail address and password, iOS users can purchase music, movies, apps, books, and more with their device.

Finally, let's talk about customer loyalty. According to the same comScore survey, nearly two-thirds of iPhone users are highly satisfied with their device, and 80 percent have previously owned an iPhone. That's a lot of repeat customers. In contrast, only 48 percent of Android device owners describe themselves as highly satisfied. It is easy to see that Apple commands greater loyalty.

What Is Native Coding?

The term "app" may have entered the mainstream lexicon only in recent years, but developers have been using it to describe their work for some time. All software, in its basest form, can be described as an app. Teenagers wrinkle their nose at stories of floppy disks and installs that take longer than a couple of minutes, but it all involves apps.

In the mobile-technology industry, a "native app" is merely an application developed using the platform's native code, downloaded from an app store, and installed on your device on an as-needed basis. Generally, native apps are no different from their Web-based counterparts. They require some sort of connection to the Internet, a database, and an interface of some kind to interact with the back end.

In the original boom of the Internet in the 1990s—often referred to as "the Dot Com era"—apps were Web-only. Page loading times were slower because of the growing availability of high-speed connections, but the concept of Web apps still applied. Simply put, each type of software has its own advantages.

Figure 7.2: iOS 7 made both positive and negative waves with design updates.

ENTERPRISE USE OF iOS CONTINUES TO GROW

Samsung unleashed a new commercial for Super Bowl 2013 that tried to change the story of Enterprise BYOD. In it, a gaming company announces it will allow employees to bring any mobile device they want and use it in the office. There are some of the usual advertising tactics, making the cool kids the one using Samsung devices. At the end of the day, the leaders of the company announce they will launch in four weeks. Guess who are the ones excited about such a short window?

Granted, this kind of commercial doesn't draw the ire it used to. Whether you are promoting a soda, a smartphone, clothing, or any other product, this is the theme your commercial will probably have at some point in time. Having played with a few of Samsung's offerings, I'm even inclined to think that a Galaxy device might be a pretty cool thing to have.

Let's not kid ourselves, however, Android is a long way from catching iOS in the realm of the enterprise.

THE REWARDS OF NATIVE CODING

When you develop apps for iOS, you will use the iOS software development kit (SDK) and Xcode, Apple's integrated development environment (IDE). Xcode provides everything you need to create great apps for iPhone, iPod Touch, and iPad. It includes a source editor, a graphical user interface editor, and many other features. Xcode employs a single window, called the workspace window, that presents most of the tools you need to develop apps. Within this window you smoothly transition from writing code to debugging

Figure 7.3: Xcode is the single entry point into Apple's "walled garden."

to designing your user interface. The iOS SDK extends the Xcode toolset to include the tools, compilers, and frameworks you need specifically for iOS.

IT'S A WILD JUNGLE OUT THERE

To get started with your first native iOS project, you will need to have Xcode downloaded and installed to your machine. If you are trying to accomplish this from a Windows machine, you will have an issue with this. Only Apple machines running OS X can download and utilize Xcode. It is also important to verify that you have the most recent version of Xcode. For every new version of each Apple operating system, there is a corresponding version of Xcode.

Once you have ensured that you have the right version, you are ready to begin! After opening the application, here is a step-by-step process for getting your first project off and running:

Create a New Project

In the welcome screen of Xcode, click "Create a new Xcode project" or just choose "File > New > New project." There will be several choices of applications to choose from. Apple has some templates of the main types of applications for both desktop and mobile platforms. You will want to make sure the "Application" section is highlighted under "iOS."

Figure 7.4: The beginning of any Apple application.

This is where the choices really start to open up for developers. You will be presented with seven different types of iOS apps to begin:

1. Master-Detail Application—it's interesting that Apple lists this application first in the list of choices, because it's the most diverse and the most challenging to use. You can specify the type of app you want to create—an app for iPhone or iPad or a universal app.

2. OpenGL Game—this is a template that will connect to Game Center. If you are superstar designer Loren Brichter, it will break Game Center because of its massive popularity. This can be pretty heavy stuff, so tread lightly.

3. Page-Based Application—a relatively new entrant in this set of templates, this attempts to have a more modern feel to the UI, incorporating multiple pages and views into the experience. This is most evident in apps like Web browsers.

4. Single-View Application—probably the most commonly selected project; use this if your app will use a single view at a time to guide the user along the desired activities you will design. It is the most versatile and the easiest to develop of the choices.

5. Tabbed Application—use this if you need an application that is similar to a website with frames. There will be a tab pane at the bottom of the app with four or five options, and choosing an option will change the activity the user experiences.

6. Utility Application—an interesting kind of application that kind of "flips." These apps tend to have a miniature "i" icon in the bottom right corner of the screen. When you click the icon, the application "flips" into another activity.

7. Empty Application—the choice for you Objective-C ninjas out there who know what you are doing with a blank slate to work with. Once you get the hang of views, models, and APIs, this is a great way to just start working on what you want right away without any of the fluff Apple adds to the other templates.

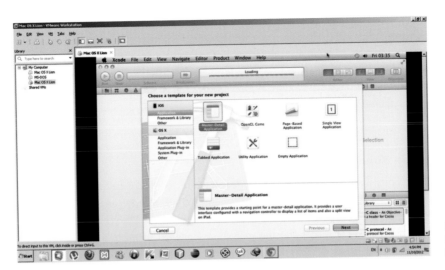

Figure 7.5: The seven templates for iOS applications.

After you select which template you need, you will be presented a dialogue box to fill in fields for your product name, company identifier, and class prefix fields. The company identifier is merely something that allows you to identify the name of the app and your company name (if you have one). It will be a necessary part of you uploading the app for approval. The naming structure is <app name>.<company name>. Think of it as a file name used to identify the app in various places. If you don't have one off the top of your head, the iOS Developer Guide recommends using "edu.self."

The next dialogue box will ask you to choose the device family the app is going to be associated with. In this box, you have the option to use Storyboard in your app development (we will say more on that in a minute) and to turn on automatic reference counting.

Finally, you will be asked to specify a location where the project will be saved. The only other option of note concerns source control. If you have a desire to document different versions of code you commit and upload, feel free to select that option. Otherwise, leave it alone for now. Once that happens, you are all set to start work, so click "Create."

Build Your Project and Run in Simulator

Even though you haven't written a single line of code yet, the template you chose (unless you selected the empty template) should have enough in it to build and run in the simulator. This is an important task to perform right away because there will be plenty of times where you don't build successfully. There's nothing wrong with this; of course, my developers break their builds on a daily basis. That said, it will be good for you to see what it looks like for the code to build successfully and run so that you will know for sure you did something wrong when things don't work in the future.

In the toolbar area in the top left corner, make sure the scheme pop-up menu has the right simulator selected (for example, current iPhone builds would need "iPhone 7.0"). Once that occurs, click the big "Run" button to build and launch. Pay attention to the status of the build because you will notice how Xcode updates you about every task along the way.

iOS GOES 64

The introduction of the iPhone 5S brought with it the world's first 64-bit-based mobile operating system. Apple is now able to cram more memory into a phone. Yes, your phone is going to be as powerful as your desktop very soon.

Remember Your Storyboard

A few versions ago, Apple introduced a new method for designing user interfaces in Xcode called Storyboards. Developers previously created XIB files for each view controller and programmed the navigation between each view manually. A Storyboard defines both view controllers (called "scenes") and the navigation between them (called a "segue") on a design surface that offers WYSIWYG editing of the application user interface. It is really handy, so make sure to enjoy this. Many developers died to bring us this information.

The first placeholder you should see in the Storyboard will be an orange cube. This is called the "first responder." It represents the object that should be the first to receive events when the app is running. The other placeholder object you will notice is called "exit"; this is for navigating segues. Think of it as a breadcrumb trail so that you can go back from the way you came. The main view controller object will be represented by a pale rectangle inside a yellow sphere. Think of it as the main scene when you are using an app. The associated view (which was the white screen you saw when first building the app) will be listed below the view controller.

There is a lot more complexity to using Storyboards, but that can be learned and built out as you define and refine the user interface of your app.

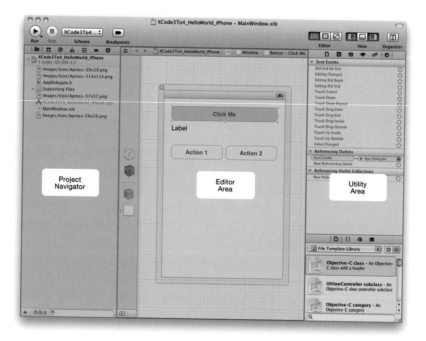

Figure 7.6: Xcode has a streamlined interface that lets you easily configure and edit your application.

Configure the View

Now you are ready to start adding objects to your main view. Xcode provides a library of different elements that can be added by dragging them to the view. Controls, text fields, and buttons are just a few of the choices you have. Storyboard allows each of the elements you display, whether they be text or buttons, to be easily configured directly in the view.

At this point, it is important to start considering the outlets for each object. The term "outlet" is used to describe a property of an object that references another object. If your view is meant to display the user's e-mail address after it has been entered and saved, you would define an outlet for a text label. Xcode defines outlets in the code with the keyword "IBOutlet." The object that defines the connection in the label is known as a "delegate."

The important thing to remember when configuring your view is that every object you add to each view needs to have a reason for being displayed. If you are adding a button to the bottom for form submission, make sure each text field in the form has a place to go and that the button has a corresponding action. Even if text is displayed to identify where you are and what you are meant to do, be sure to identify the connection and delegate.

Don't Forget the View Controller

The importance of the view controller can be stated no better than what is written in the iOS Developer Library: "View controllers are a vital link between an app's data and its visual appearance. Whenever an iOS app displays a user interface, the displayed content is managed by a view controller or a group of view controllers coordinating with each other."

Simply put, the controller is the skeleton that props up your view. How users switch between views, access data from tab bars, and store data entered in text fields are all defined here. There can be a controller for the main and secondary navigation, as well as for every single page of content.

Views and controllers are two of the three functional aspects of the MVC development framework for software. The third aspect is the model, which encapsulates the data specific to an application as well as the logic and computation that manipulates data. The view is the common object in the center, and models and controllers act on either end to perform your necessary functions.

Before you set out to start defining controllers and using them to connect a set of views, remember what a storyboard is used for in commercials: to tell a story. Even if you need to define how users will navigate from screen to screen using sticky notes on a giant wall, map everything out in as much detail as is needed. I can't tell you the number of times a developer has wished he had done the same thing when he forgets a vital view that isn't hooked into the flow.

Once your data and interactions are all defined and flowing, you can definitely start to see your app taking shape.

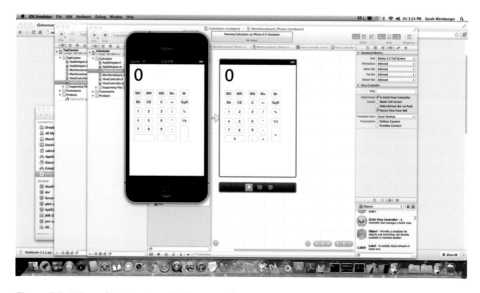

Figure 7.7: iOS applications heavily focus on the integration between Views and Controllers.

What Do You Mean I'm Not Done?

Code review is one of the most dreaded aspects of app development, but it is very important to make sure the code compiles without any warnings in Xcode. There are errors, which means the code did not compile correctly (which is pretty much a showstopper) and warnings. Simply put, your code may compile and run, but there might be some issues that keep your app from executing to its greatest potential. Apple recommends that all warnings be resolved, because in many cases that is all you will see.

The sources of warnings are often improper or missing connections. If text does not update when you touch the button or the keyboard does not disappear when you submit, check the connection first. If the delegate is properly connected, make sure your delegate method is spelled properly. To ensure this, many developers just copy and paste all declarations.

Here is an example of an interface and implementation file from a sample app in the iOS Developer Library called "HelloWorld":

The Interface file:

```
#import <UIKit/UIKit.h>
@interface HelloWorldViewController : UIViewController
<UITextFieldDelegate>
@property (copy, nonatomic) NSString *userName;
@end
```

The Implementation File:

```
#import "HelloWorldViewController.h"
@interface HelloWorldViewController ()
@property (weak, nonatomic) IBOutlet UITextField *textField;
@property (weak, nonatomic) IBOutlet UILabel *label;
- (IBAction)changeGreeting:(id)sender;
@end
@implementation HelloWorldViewController
- (void)viewDidLoad
{
[super viewDidLoad];
// Do any additional setup after loading the view,
typically from a nib.
}
- (BOOL)shouldAutorotateToInterfaceOrientation:(UIInterfaceOr
ientation)interfaceOrientation
{
return (interfaceOrientation ! =
UIInterfaceOrientationPortraitUpsideDown);
}
- (IBAction)changeGreeting:(id)sender {
self.userName = self.textField.text;
NSString *nameString = self.userName;
if ([nameString length] = = 0) {
nameString = @"World";
}
NSString *greeting = [[NSString alloc]
initWithFormat:@"Hello, %@!", nameString];
self.label.text = greeting;
}
- (BOOL)textFieldShouldReturn:(UITextField *)theTextField {
if (theTextField = = self.textField) {
[theTextField resignFirstResponder];
}
return YES;
}
@end
```

Courting Feedback with Some Easy Tools

We are too close to our own work. It's where the phrase "eating your own dog food" comes from. We spend so much time with our idea and execution that we forget to look objectively at each and every aspect of the project. The sooner you can start getting your app into the hands of outside users, the sooner you can start learning what you failed at and improve immediately.

Apple has provided an easy tool, AdHoc, for doing beta testing. The tool can be used to set up a quick distribution system to other iOS device owners who wish to check your creation out. When your code is in a stable enough state, choose "Build and Archive" from the "Build" drop-down. Once that happens, the Organizer window will appear (also found by selecting "Organizer" from the "Window" drop-down). Select "Archived Applications," and make sure the build you just ran displays.

A list of sharing options will appear. After clicking "Share Application," click "Distribute for Enterprise." Then fill in the app URL, title, subtitle (optional), and image URLs (optional). Click "OK," and specify where the file should be saved. Make sure it has the same name you provided in the URL.

Finally, make sure you upload all the correct files. The .plist and .ipa files generated by the build will be first; then you need to create a provisioning profile and simple index file. Here is a sample:

```html
<!DOCTYPE HTML PUBLIC "-//W3C//DTD HTML 4.01 Transitional//
EN" "www.w3.org/TR/html4/loose.dtd">

<html>

<head>

<title>My Cool app</title>

<!—Art Direction Styles—>

</head>

<body>

<ul>

<li><a href = "http://jeffreysambells.com/example.
mobileprovision">

Install Example Provisioning File</a></li>

<li><a href = "itms-services://?action = download-
manifest&url = http://jeffreysambells.com/example.plist">

Install Example Application</a></li>

</ul>

</body>

</html>
```

The .plist file also needs to include your bundle ID. You set it manually or find it in the "Targets" section of the project navigator. On the "Info" tab, enter "App ID" in the "Value" column of the "Bundle identifier" row.

Once this is done, all you need is to point users to the index file, and they can click the links to install the provisioning profile and start using the app. Some commercial solutions to this very task have gained in popularity in the past year. Try TestFlight and Hockey if you want a more "off-the-shelf" experience for providing a file to beta testers.

A NOTE ABOUT REJECTIONS

Google made the kind of news in April 2013 that rarely appeared in the past, concerning app rejections. Now that services such as Applause can actually document that the quality of apps is higher for iOS than for Android devices, looks like the Android owner decided to do something about it.

Standards should be kept high, and with iOS you can be sure the platform has been upholding this standard from the very beginning.

Managing Your App Using Xcode

If what we have described were all Xcode provided to iOS developers, it would be a fantastic tool that allows even the most novice users to quickly start testing out new ideas. There are, however, some useful ways to inspect and adapt your existing code base as new technologies and user experience trends emerge.

MANAGING SOURCE CONTROL

You may remember the subject of source control management from the step-by-step instructions. There are two types of SCM systems supported in Xcode: Git and Subversion. Both are two of the top development tools in the community and offer easy solutions to allow companies of all sizes the ability to host and manage code bases.

A version editor is included to make it easy to compare versions of files saved in code repositories from different systems. If you have introduced bugs to your code, this is an easy way to compare changes between versions and quickly zero in on the source of the problem.

In the "File" menu option, you can choose to create a snapshot of your code. If you remember to do this regularly in the development process, you will have a tangible set of progression to inspect. In the Organizer window of Xcode, if you click the "Projects" button, the screen will display a list of all the snapshots taken. This allows you to restore an old set of the code in case a reversion is needed.

IMPROVING PERFORMANCE

Perception is always a priority in software. When the iPhone was released, perception became even more important to developers because how your app is perceived to run is much more different than how it actually performs. When you "like" something on Facebook's app, the action of liking something is not as immediate as the touch leads you to believe. It is merely queued in the back end of the software and most likely will happen in a matter of minutes (depending on current usage volume).

Some of this relates to how you design the completion of every interaction with users. If the app runs slowly, that almost guarantees that your app will be deleted as quickly as it was downloaded. The simulator can be a bigger help than many would think, but it can never replace actual usage in a device. Xcode has a tool named Instruments that gathers data by running your app and presents it in a graphical timeline.

Information about the app's memory usage, disk activity, graphics performance, and more can be measured using Instruments. Viewing the data can show you where hiccups in the code are occurring. The tool can also automate the testing of interface elements in the app.

To get started, choose "Product > Perform Action > Profile Without Building" in the project section of your app. There will be a button in the left column named "All" that can display all the trace templates once clicked. Once you have selected all of the data points you want to measure, click "Profile," and the Instruments app will run along with Simulator.

As you interface with your app, notice how the data display in Instruments. This can be most effective when performing end-to-end flows like account creation. Keep an eye out for large spikes in any of the tracking graphs, as this can mean that an undesired side effect is occurring in the code. Bottlenecks in performance might mean that you need to pre-allocate some blocks or lazy-load others.

RUNNING YOUR APP WITH SIMULATOR VERSUS AN ACTUAL DEVICE

Since it has already been mentioned, let's cover how to install your app on an iOS device for what many call user-acceptance testing. In the Organizer window, choose "Devices." Under "Library," select "Provisioning Profiles." If an option to enter your Apple Developer user name and password does not appear, click "Refresh" at the bottom of the window. Then you can log in. The tool will ask whether Xcode should request your development certificate. It will need to be added to your keychain and, later, to the iOS Team Provisioning Profile.

To run the app on a device, install the associated profile on the device. This enables your app to run by identifying the device to Xcode. If the device is provisioned properly and is connected to your Mac, it will appear in the scheme editor. If you select it in the Destination pop-up menu and click "Run," you will see the app appear on your home screen.

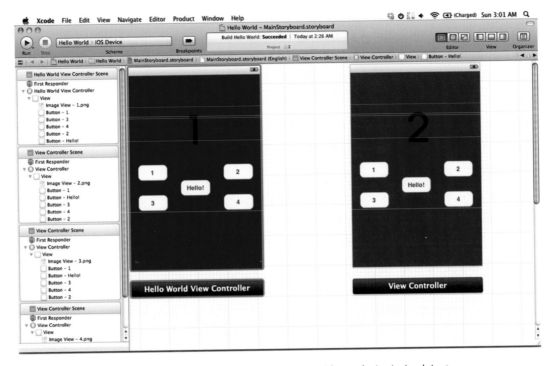

Figure 7.8: Build your app using the simulator for easy testing with any device in Apple's store.

iOS REMEMBERS ITS OWN

Bloggers have written for a few years about the top "replacement" apps, and I don't blame them. With the release of iOS 7, the biggest win might just be the highlight of the native apps discarded for years. Now, users can feel more comfortable about deleting these replacements and giving Notes, Reminders, Weather, and Safari a try again.

iOS Development with Objective-C: Speed, Speed, Speed

Even though Xcode does some of the work for you, it is important to understand the fundamentals of the primary programming language, Objective-C. It is a simple, yet sophisticated, object-oriented language that extends the standard ANSI C language by providing syntax for defining classes and methods. It also promotes dynamic extension of classes and interfaces that any class can adopt. As a superset of the C programming language, you get all of the familiar elements, including primitive types, structures, functions, pointers, and control-flow constructs.

Here is a list of the syntax that Objective-C adds to ANSI C:

1.   Definition of new classes. There are two distinct pieces that encapsulate a class in Objective-C: the interface and the implementation. The interface includes the class declarations and defines the public interface of the class.

2.   Class and instance methods.

3.   Method invocation (called messaging).

4.   Declaration of properties (and automatic synthesizing of accessor methods from them).

5.   Static and dynamic typing.

6.   Blocks (encapsulated segments of code that can be executed at any time).

7.   Extensions to the base language such as protocols and categories.

For those new to object-oriented programming, think of a method as a function that is scoped to a specific object. By messaging the object, you call the method. The two kinds of methods in Objective-C are instance and class. Instance methods refer to a specific instance of the class, which has to be created before you can call it. Class methods don't need this definition because they refer to the entire class. The declaration consists of an identifier, return type, signature keyword (or keywords), parameter type, and name information:

```
-    (void)  insertObject: (id)anObject atIndex: (NSUInterger)
index
```

To dispatch a message, the runtime requires an expression, which encloses brackets and the message itself (along with parameters) and the object receiving the message. Remember that you can also send messages to the class itself.

Properties are declared at the beginning of the method in your class interface using the designation "@property." The header files usually contain them. Basic properties include the type of information and the name of the property. Custom options also follow, contained in parentheses. Here are some examples:

@property (copy) MyModelObject *theObject;   // Copy the object during assignment.

@property (readonly) NSView *rootView;   // Declare only a getter method.

@property (weak) id delegate;   // Declare delegate as a weak reference

Blocks are objects that encapsulate a segment of code. They are portable and anonymous functions that can pass in as parameters of parameters of methods and functions. Blocks may have an inferred or a declared return type. It's also possible to assign a block to a variable and call it just like a function. The synthetic market for blocks is the caret (^):

```
Int  (^myBlock)  (int)  =  ^(int num)  {  return  num  *
multiplier;  };
```

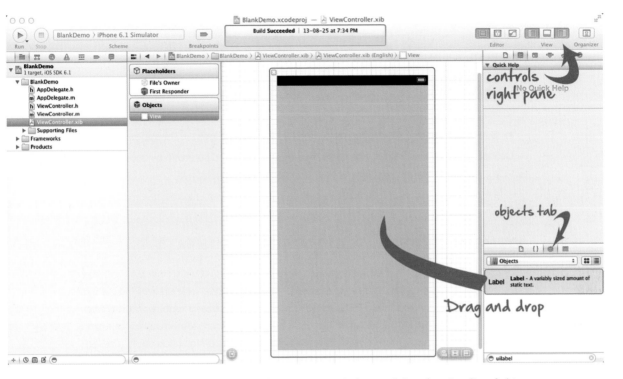

Figure 7.9: Xcode allows developers to work through their process with drag-and-drop functionality of objects.

Protocols declare methods that can be implemented by any class, even if the classes implementing the protocol don't have a common superclass. The beauty of protocols is that they are independent of any particular class. They simply define a list of methods that connect objects without needing them to be specific instances of a class. Instead of creating a subclass to receive a notification necessary for a class, you can use a protocol to call a specific method and deliver the notification and provide the necessary response.

In a block, specify that your class adopts a protocol by putting its name in angle brackets (< . . . >) after the name of the class in which it inherits. Keep in mind that you don't need to declare protocols:

```
@interface HelloWorldViewController : UIViewController
<UITextFieldDelegate> {
```

Finally, let's define some common terms used in Objective-C:

- Id: The dynamic object type. The negative literal for both dynamically and statically typed objects is "Nil."

- Class: The dynamic class type. Its negative literal is "Nil."

- SEL: The data type (typedef) of a selector. It represents a method signature at run-time. Its negative literal is "NULL."

- BOOL: A Boolean type. The literal values are "YES" and "NO."

- Self: A local variable that you can use within a message implementation to refer to an object. The C++ equivalent is "this." When messaged, the runtime looks in the method implementation in the current object's class.

- Super: Similar to "self" with the exception that the runtime looks in the superclass first.

Defined typed and literals are often used in error-checking and control-flow code. You can test the appropriate literal to determine how to proceed in the statement.

OTHER LOCATOR SERVICES MAY SURFACE SOON IN iOS

In March 2013, Apple purchased the indoor location startup WifiSlam. This industry is not an unknown to Google, which already employs it in many countries in public locations. If Apple could get a step ahead of its main competitor in this space, however, think how the conversation could change with regard to mobile applications.

WifiSlam uses wifi signals to position devices within a 2.5-meter radius of your location. Just think how that could change app development in the next few years. Instead of your using your device merely to locate a shopping mall, you could use it to locate your favorite store once you are inside the mall.

Native Kits for Rapid iOS Development

APPKIT VERSUS UIKIT

You will read a lot online about the strengths of both of these libraries, but it is important to understand each individually and how it applies to Apple's two platforms.

The AppKit is a set of classes, first developed by NeXT Computer and now by AppleComputer, for rapidly developing applications. The majority of AppKit classes deal with user interface elements, such as windows, buttons, and menus. InterfaceBuilder lets you work with many of these classes visually. AppKit is from an older time when RAM was scarce, CPUs were slow, and GPUs didn't come in consumer machines.

UIKit is a relatively new framework created solely for iOS. Look at it as the company's chance for a clean slate with a fresh library of tools engineers can use. The Core Animation method, which was light-years ahead of any competition when it first was released, allows for simultaneous manipulation of layers. Any game that doesn't utilize this technology loses significant quality.

MAPKIT

This framework provides an interface for embedding maps directly into your own windows and views. It also provides support for annotating the map, adding overlays, and performing reverse-geocoding lookups to determine the place-mark information for a given map coordinate. One important note here is the change iOS 6 brought to this library regarding native functionality. Originally, Google Maps was the native application used by this kit. Now, with Apple still trying to establish its own map application in the market, all class and protocol references route users to Apple Maps.

While Apple has taken its lumps over the fumble in launching the software in 2012, a lot of work has been done on it, and engineers are taking note.

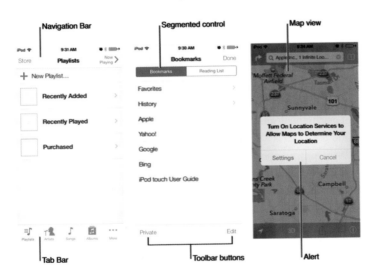

Figure 7.10: Utilize native functionality of iOS devices to provide a rich experience for users.

GAMEKIT

When the iPhone was first introduced, nobody could see how it would evolve into one of the most popular game platforms in existence. All those who own or know someone who owns one of the devices can name their favorite game that gained prominence through the iOS platform. One aspect of gaming that grows in popularity every year is the social context of playing a game.

Sharing a score is now much more than posting about it on Facebook. Game Center's ability to connect users worldwide accomplishes that. While it was once a luxury to have connectivity to the iOS functionality, it is now a requirement if you want to keep up with talented studios like Rovio and PopCap. GameKit provides that with leaderboard updates, social networking, voice-over IP, and more.

Submitting Your App

GETTING YOUR DEVELOPER LICENSE

You do not need to enroll in the program to write apps and test them in iOS Simulator. You do need to enroll, however, to test apps on devices and to distribute apps. Enrollment gives you full access to the iOS Dev Center and the iOS Provisioning Portal. If you enroll now, you will be able to follow all of the steps in the road map, including testing your app on a device.

Figure 7.11: iTunes Connect easily connects your source code to Apple for submission.

The two types of developer accounts are individual and company. The main reasons for designating your account as company are the ability to have multiple users on the team and the fact that it allows you to get a DUNs Number. The Data Universal Numbering System is a copyrighted, unique identifier for your company with Dun & Bradstreet that unlocks a wealth of value-added data. All companies doing e-commerce use this number. For more information or to register, you can go to www.dnb.com/get-a-duns-number.html.

Once you have registered your account at developer.apple.com, you can use the Member Center to manage program accounts, register App IDs and devices, create signing certificates, and create provisioning profiles. You can also use it to access iTunes Connect.

Beginning your submission starts with creating the app in iTunes Connect. You will need to create a project here much the way you did in Xcode. Select "Manage Your Applications" and click "Add New App."

NEW APP DATA

Regardless of your social media following, your marketing plan, or the line of rabid fans waiting for your next offering to the iOS App Store, the one place you have the same chance as everyone else to brag about your product is the listing for your app. The name, description, icon, screenshots, and contact information you provide can be the difference between someone giving your software a try and moving to the next app. Your record will also need the date when you want your app to be made available. If you aren't sure, you can always set the latest date possible and change it later. Also, remember the bundle ID created when you first started the project? Make sure it matches what you enter here.

The iOS Developer Library provides precise information about the resolution for icons. It is important to fill in every requirement because you will get the quickest rejection ever if you don't provide this information. In that same vein, don't forget to show off your app in the best possible way in the screenshots you send. If you don't have user reviews (which many times can be more valuable than the marketing data in the App Store), these screenshots will be the only way for a potential user to see the UI for your software.

Figure 7.12: The better you describe your application in iTunes, the easier it will be for users to make the decision to download.

Many times, companies update screenshots whenever updates to the software are made to let users know of new enhancements. If users are familiar enough with the original screen shots, this can also be a way to attract increased usage once they put it away for a while. The last piece before submitting the code for approval is to upload the binary and fill in the compliance question. Fortunately, the software does all of the work for you, provided you have done everything correctly in Xcode. Once this is done, your app will have the status "Waiting to Upload."

The archive that you submit must be signed with your distribution certificate. This was created in the provisioning process of device testing. Apple does provide some simple validation testing to check the archive against information you provided in iTunes Connect. Of course, being the great engineer you are, you are performing these checks anyway during the development process. After the archive is created and validated, you are finally ready to submit and ship your app.

Xcode validates the archive again during the submission process and then submits it for approval from Apple. You can sometimes hear back in as little as a few business days, but with the popularity of the store you can sometimes expect to wait quite a bit longer. I have heard of some devs waiting for weeks, which calls to attention the importance of testing and validation to ensure you don't get a rejection.

While we are on the subject, don't get hurt if your first submission gets that designation. My first rejection was the result of my not including some simple "how to" text in the sign-on screen to help direct users to the right actions. Usually, the information provided by Apple is sufficient, but the customer support with Apple is top notch, so reps will be more than happy to walk you through the issue if need be.

KEEP AN EYE ON iTUNES RADIO

In a June 2013 article, the *Guardian*'s Charles Arthur argues that the rate Apple is going to pay artists for play in iTunes Radio could potentially make the service very attractive to record labels.

It's no secret record labels love and hate the team in Cupertino because of how iTunes affected music sales, but it gets harder to make money from free streaming music every year. That's where the cash Apple has on hand comes in handy. Arthur states: "What the other streaming services have discovered repeatedly is that it's hard to make such a service profitable, because the music costs don't fall as they grow—in web terms, it doesn't 'scale'. Thus Spotify has put a 10-hours-per-month ceiling on free listening, and Pandora blocks people outside the US from listening."

So, while labels have been publicly supporting Spotify for some time, it will be difficult for labels to not throw their good stuff to Apple when it pays ten times more. Add to it the easy integration with iTunes to purchase music, and this looks to be an easy win for iOS. It's crazy to think that the announcement was merely a footnote in the WWDC keynote presentation.

How Can You Make Your iOS App Stand Out?

I remember reading once that what you say is more important than how you say it. While that can apply to product vision and marketing of your software, the most important thing you can do in your app is to watch "how" your app does things.

Remember that gestures are valuable. I'm sure all of you noticed how the left and right swipe has replaced back and forward buttons in the latest version of iOS. The UI libraries can take advantage of any kind of gesture you can think of. Touches, swipes, drags, pinches, even blowing into the microphone can help users interact with your code in unique ways.

Be sure to also analyze each screen and maximize every pixel. There's no need to waste space on navigation bars as much as before because of the previously mentioned swipes. You can hide navigation on either side and save your real estate for the good stuff. You can also wow Apple device users with a clean interface that puts design first.

Finally, don't make your users' fingers go numb before an end-to-end loop is complete. Measure the time and number of touches it takes to complete an order, register for an account, or create a character. Nothing makes a user close and delete your app like becoming frustrated before getting into the meat of your experience.

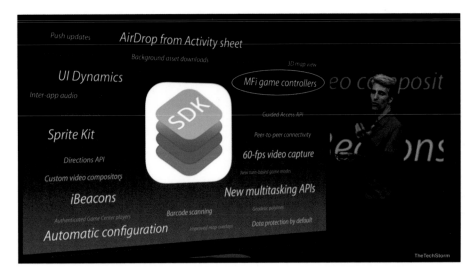

Figure 7.13: The number of enhancements to iOS 7 gives developers even more tools to work with.

CHAPTER 8

Developing for Android

There will always be two sides to the argument regarding Android versus Apple. While Apple focused on making its margins from the hardware that houses the apps we develop, developers for Google's Android OS typically focus on services first. All an OEM would need to do is to purchase Android licenses for its plethora of devices, and it could make it its own with Android's open standard. This has led to a predictable result.

Figure 8.1: KitKat was announced in September 2013 to the delight of Nestlé and candy lovers everywhere.

Android is the leading mobile operating system in the industry, and its margin is only growing. The greatest example is in a realm that has been mostly owned by iOS: tablets. According to survey results from Strategy Analytics in July 2013, Android has gone from being essentially tied for market share with Apple's OS (at 51 percent to 47 percent) to enjoying an almost 3:1 advantage (67 percent to 28 percent).

When you factor in individual OEMs, the race for market share becomes more favorable to iOS. You don't have to worry about this from a development perspective. You just care about your app being used by the most people possible.

Myths and Facts about Android

Although Android was once the media's darling, some of the original myths about it have started being debunked. This has happened slowly, over time, because the OS did not come out of the box polished (not that iOS did, either). What Google has done, however, is to iterate quickly and take note of the community's thoughts. Even though the first Android-powered phone was delivered in October 2008, the software is on its tenth major release (18 releases total including all the various API levels). For comparison, iOS is on its seventh.

In his March 2013 mega-opus on why he switched from iOS to Android, tech journalist Andy Ihnatko summarized perfectly the thoughts of many mobile users today: "[When people asked,] I would say, 'If someday in the future somebody makes a phone and an OS that's a better fit for me and my peculiar needs than the iPhone, I'll make the exact same choice.' Yep: that day has come."

While many of the high-end development houses still release to iOS first, many are realizing that it is worth the effort to port over to Android. With more than a million available applications already in Google Play, developers are stepping into these waters first. Depending on whom you talk to, there are advantages to dual-platform releases and staggered releases. There are advantages to free, "freemium," and paid downloads. It would be a mistake to assert any statistics as standard for the entire industry, because category and audience are huge factors to weigh.

It would be a huge mistake to think that Android users are less willing to pay for software they download, even if it was the case before. Robot Invader CEO Chris Pruett gave a talk comparing iOS and Android games at the Game Developers Conference in March 2013 in which he discussed this very fact. He stated that for the Android and iOS versions, the conversion rate of users making in-app purchases was virtually the same (less than a percentage point of variance between the two platforms).

What Pruett did mention as a huge advantage of the iOS version was time of play. Many studies have shown that iOS device users spend much more time on their phones or tablets than Android users, and this was no different for Robot Invader. Depending on the specific version of a game, the time of use was two to four times greater for those on Apple devices.

Figure 8.2: Google provides developers with a wide array of tools to give users the best mobile experience possible.

One important distinction between iOS and Android platforms is support. Apple charges more for developer licenses but does its best to own the first line of defense in support of software it releases. With Google, the developer is responsible for every step of supporting its users. I don't feel that this is a huge disadvantage, but it is something to consider when setting up your product model.

Language support is also discussed heavily in mobile development because users are literally all over the world. Some would argue that the more languages the better, but again this points to the importance of knowing your target audience. Many developers still focus on English and one other language, such as Spanish, Japanese, or Chinese, to stay diverse. Depending on the size of your company, you can't expect to support every language, so pick your battles.

The biggest myth that still gains traction to this day is the fragmentation of the mobile platform. Because of its open-source capabilities, there are hundreds of devices and distribution models to worry about. As the past couple of years have shown, however, there are really only a few heavy hitters in the cutthroat world of Android OEMs. Samsung, HTC, and Motorola are all vying for shelf time next to Google's own releases.

You can take two big steps to help combat this challenge for your product: do your research and plan well. Making sure you have a fluid design that allows for multiple screen resolutions and densities will go a long way in allowing for as many users as possible. It also does a bit of future-proofing for devices you don't yet know about. Also, keep an eye out for sales numbers for devices. While you have to be careful to know the difference between distribution and sales numbers, there is plenty of data available to guide you in deciding which should be your top five choices of devices. If you can do both of those things regularly during the planning and development stages, you will be a step ahead of your competitors.

MORE AND MORE ARE CHOOSING ANDROID

Tech journalist Andy Ihnatko, who has been a huge Apple fan for years, made huge waves in early 2013 by making the switch from iOS to Android. The Chicago *Sun Times* writer has a reputation for integrity and for his love of Cupertino products. Needless to say, it was a big deal when he switched.

It's popular to be dumping on Apple now. Around the same time, tech blog GigaOm posted a podcast that contained nothing but Apple users calling in to say they are dumping the iPhone. Social media have allowed the Samsung versus Apple debate to reach a fever pitch.

The piece by Ihnatko goes into great detail about his reasons for switching, but the reasons are different for many users. At heart it all remains the same: people want something new, and they are finally willing to switch platforms to get it.

Know Your Design Elements

Before you get started, there are a few common terms that are used in the Android platform that you would do well to get acquainted with. These elements are often used when discussing different trends and issues in the Android developer community. These design options can add depth and functionality to your software that only the Android platform provides.

1. Tabs—this UI element is used heavily in the provided templates by Android, used most often in swipe views. This is usually a part of action bars to allow for easy navigation. Tabs can be scrollable, fixed, or stacked.

2. Lists—one of the more popular elements, lists let you show multiple items in a vertical arrangement for navigation or display of data. List can be utilized with one-, two-, or three-line arrangements.

3. Grid Lists—when lists don't work for your design, grids are the next logical way of displaying text or images. The grids utilize either horizontal or vertical scrolling elements to help users navigate through the list.

4. Scrolling—unless your design has a fixed, full-size display, at some point you will need users to scroll. When you do, make sure your scroll indicator is properly displayed.

5. Spinners—also known as pickers, this element is a section method for data types or navigation that is denoted by a triangle in the lower right corner. Touching a

spinner displays choices in a dropdown menu. Note that pickers need not necessarily use the triangle.

6. Buttons—the most common method for denoting something you want a user to touch, buttons are displayed in various forms, the two most common are shapes and borderless icons.

7. Text Fields—used for showing users where you want data entered. Android provides the ability to capture information in single-line and multiline fields. One thing to remember is that you can easily add auto-complete functionality with this platform.

8. Sliders—most common in settings menus for setting values with a single gesture.

9. Progress and Activity—these are indicators used in multiple interfaces to show how far a user has progressed with a flow. The two most common methods of implementation are bars and circles. Newer trends are also making room for "pie chart" styles of indicator.

10. Switches—also used most often in settings menus, this item is for the selection or deselection of something. The three methods most used are checkboxes, radio buttons, and on/off switches.

11. Dialogs—these boxes were invented by Web developers and can be very valuable for communication between your app and users. Variations of this element include alerts, pop-ups, and toasts. Almost all versions of dialog boxes include a title, a content area, and an action button of some type.

Figure 8.3: The design possibilities are endless with Android.

Utilizing Android's Simple Workflow

Google makes the steps to publishing your app in the Google Play store as easy as possible. Instead of worrying about hoops to jump through, you can focus on giving your users the best work imaginable. Android employs the "many paths up the same mountain" strategy, because engineers develop software in different ways. Why make them fit into a box when doing so might limit creativity?

This section addresses execution of the four stages of development: setup, development, debugging and testing, and publishing. Once you have the initial flow down, you can iterate with ease to maximize your output.

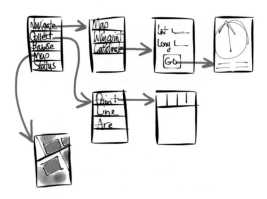

As always, let's get started with tools first. The SDK can be found at developer.android.com/sdk and provide all the proper links for your machine. The initial download includes access to Android Developer Tools (ADT). The ADT bundle gives you access to the latest version of Eclipse Integrated Developer Environment (IDE) and an ADT plugin, SDK tools, platform tools, the latest version of Android, and the latest version of the emulator.

While Eclipse is included in the download, you are not required to use it for your workflow. Other IDEs are allowed. You just need to make sure the ADT plugin is correctly installed and pointed to the SDK. Once this is complete, you are ready to begin your work.

Figure 8.4: Storyboarding the flow of your application can avoid development mishaps.

APPS THAT REAL PEOPLE USE

In a March 2013 article, *Forbes* broke some impressive ground for readers. Apparently, the highest-rated apps, according to current analytics, are pieces of software that have features people will actually use. Of course, that brings up the relevant point: just how exactly do you know which features people want on their mobile devices?

The short answer is, it depends. Since this article covers in the retail industry, let's use that as an example. Mobile devices are meant to connect people who would otherwise not be connected. For retailers, usually that means getting people into the store. If you can utilize some sort of digital retailing system or rewards based upon location, this may help drive traffic to the building.

The days of half-built mobile platforms are gone. If you don't want to end up like the infamous examples in this article, know your brand and increase its awareness with an app people will love telling their friends about!

Setup

While it involves more than just creating a developer account, setup on your machine requires merely downloading and configuring the methods to test your application.

The first thing is to have an account with Google Play. Having a Google account already takes care of much of this process, because your information will already be in the database.

If you visit play.google.com/apps/publish/signup and sign in, the rest of your developer account configuration involves three steps:

1. Acknowledge the distribution agreement—it's a basic terms-and-agreements document. Make sure you understand the expectations of Google and you should have no problems.

2. Pay the fee—it's only $25, which is one-quarter Apple's account fee. It gives you access to the SDK and all the bells and whistles involved.

3. Make sure your merchant account is attached—we all want to get paid for our work. Android requires all of the same information covered in chapter 7, along with setting up the countries of distribution.

Use of Physical and Virtual Android Devices

Before you start work on your application, it is vital that you set up all the methods of testing it with Android devices. As with iOS, there are physical and virtual methods. Knowing the advantages of each can help you ready your app for distribution as soon as possible.

Physical devices give you access to all of the sensors installed by the manufacturer, such as the Accelerometer. It also can provide an experience similar to what your users will have when they first download the device. You can also test away from your machine if you want to use GPS libraries. While Google recommends a Nexus device for testing without the need for a SIM card, any device is okay.

One important note: make sure you know the version of Android before you purchase a device for testing and debugging. Older models may be cheaper, but they may also have a version of the OS that hinders development. I speak from experience when I say that nothing is more frustrating than testing your app on a device that many of your users won't buy. Do your homework first.

The SDK allows for two methods of publishing your work to a physical device: direct installation from Eclipse and command line installation with Android Debug Bridge (ADB). While the SDK provides simple instructions for both methods, there are three things you need to keep in mind for setup of each device:

Figure 8.5: The countless number of Android devices requires you set up a wide variety of virtual devices for testing.

1. Declare your device in the manifest—this ensures that you can utilize debugging.

2. Enable USB debugging—this can help you log your progress through the testing phase.

3. Set up your system to detect your device—you don't want to have to go through setup every time you connect.

Virtual devices are much easier to set up and don't require you to connect anything before you start debugging. You can determine the device parameters you want to use beforehand to provide a more thorough battery of tests, as well as ensure compatibility with as many devices as you wish. While it is connected to your machine and doesn't involve all of the device sensors, most of that can be simulated as well.

The tools around Android Virtual Devices (AVD) can be accessed from Eclipse or from the command line in the <sdk>/tools directory. The three step of setting up each AVD are: creating a hardware profile, mapping the profile to a system image, and creating a virtual storage area on your machine for each AVD. Create as many of the devices as needed before you start work so that compiling and testing are a seamless exercise.

As with physical devices, it is important to know to which API level you are mapping the AVD. Do your research on the devices you wish to distribute your software to, and make sure the right version matches. It can be very frustrating to deliver a build to users that won't work the way you meant it to. As well, it is a good idea to develop with some forward compatibility. While Android OEMs may not always be up to date with the latest version of the OS, at some point they will release an upgrade for the device. When that occurs, make sure you are ready because they don't always do a good job of letting the development community know when this is happening, as Apple does.

BE WARY OF NFC INTEGRATION

Android users love to tout their devices' near field communication (NFC) capabilities. At the Mobile World Congress in February 2013, a slew of NFC-enabled phones were unveiled. Many current Android devices include the chip.

At MWC, a partnership was also announced between Visa and Samsung to bring mobile payment software to those devices. On the surface, it's great. Good for Visa to get out in front of the pack. The attempt to standardize software for mobile payments could be what the market needs to help bring mobile payments to the masses. Unfortunately, it's a little overly optimistic at this point.

The big chains will most likely adopt this new technology much sooner (if they haven't already), but with the advent of Amazon those stores are starting to die. Smaller retailers will be the decider.

Where I see the most value is with the transactions we all make, such as purchasing gas. Of course, there is very little difference in opening my phone and waving it in front of the pump and pulling my credit card out and swiping it. Market research even shows US customers still prefer plastic at this point.

NFC may eventually become the standard, but Apple is doing everything it can to keep that from happening at the moment. If Android users are your sole audience, then fire away with integration. If multiple platforms are your end goal, though, be extra careful because the future of the technology isn't decided yet.

Development in Eclipse

Some of you may be old school with your development process, but Google has provided the simplest way to manage your project in Eclipse. The graphical user interfaces (GUIs) and wizards associated can quickly complete a lot of the creation process for the three versions of a project:

1. Android projects—this is the base project. It incorporates all of the resources necessary to produce an .apk file, which is what Google Play needs for distribution.

2. Library projects—this comes in handy when you want to use your initial Android project as a library for other applications. This is how you turn your software into a platform. Facebook, Evernote, and Salesforce were once just applications to solve a problem. Now, developers worldwide incorporate their APIs and libraries into their projects.

3. Test projects—most useful during the testing and debugging stage, covered later, this type of project extends JUnit functionality into the Android platform. JUnit is an open-source framework for the testing of Java-based code. Repeatable test methods (@test) can be used to incorporate test-driven development into your process.

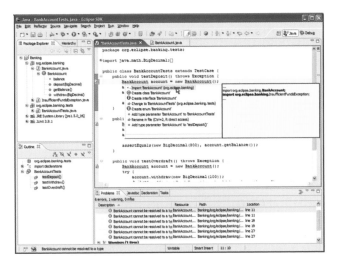

Figure 8.6: The most popular IDE for development currently is Eclipse. Android provides easy integration with its SDK and Eclipse.

One important note: all projects must start out as Android projects. The other two versions can be converted afterward.

Setup of an Android project can be accomplished with the wizard in Eclipse through "File > New > New Project." Then select "Android > Android project" and click "Next." After that, enter some basic setting parameters:

*   Application name
*   Project name
*   Package name
*   A minimum required SDK
*   A target SDK

- A "compile with" API level

- A theme

- An icon

The Activity Life Cycle

You will read a lot online about the purpose and use of the "activity" in Android, so it is important to define and know the entire life cycle. The term is just another way of describing an interaction the user is doing, has done, or will do. When an activity first starts, it comes to the foreground for the user to interact with. As the activity becomes secondary, a series of life cycle methods transition it to the background and ready it for use again. Some activities are meant to run in the background as well, so it doesn't necessarily need to be primary to a flow for it to be considered so. As such, it is important to define how the callback method behaves when it leaves and re-enters.

Other development frameworks may define launching apps in the "main()" method, but not Android. Here, you use specific callback methods that correspond with specific states in the activity life cycle:

- Created, a new activity that has not been started: onCreate()

- Started, which means it's visible: onStart()

- Resumed, visible state that was paused: onResume()

- Paused, a partially visible state that is obscured by another activity: onPause()

- Stopped, different from paused because it is hidden and not running: onStop()

- Destroyed, which can't be resumed: onDestroy()

Note: not every state in the life cycle is needed for every activity, so don't worry about defining every method that way.

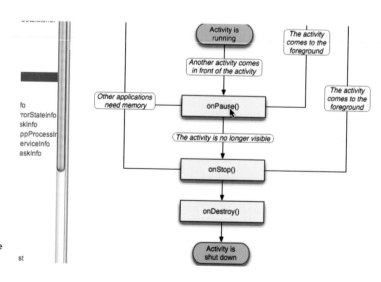

Figure 8.7:
Keeping track of
your application's
activities allows
for better resource
management.

Selecting the Right Code Templates

The final elements needed are code and activity templates. Knowing the right template for your project can make a difference in the amount of growing pains of your first effort. As with iOS, the templates provide the necessary file structure to at least run and build your app successfully before injecting your own code variations. Application templates can show the structure for the entire build, activity templates can add new wrinkles into your existing template, and object templates can inject outside integration into your app.

Three application templates currently exist:

1. Blank Activity—this is the most basic template because it does not include much additional code other than the basic structure. However, you can customize the template further by selecting the type of navigation structure you desire.

    i. None: includes a title bar, options menu, and basic layout according to the Android design guidelines.

    ii. Tabs/Tabs + Swipe: includes an ActionBar for tab controls, fragment controls for section content, and optional swipe gesture support.

    iii. Swipe Views + Title Strip: includes PagerTitleStrip for section titles, fragment object for section content, and a FragmentPagerAdapter to manage section fragments.

    iv. Dropdown: includes an ActionBar for list node navigation and fragment objects.

2. FullScreenActivity—this template is most useful for apps that want to alternate between the standard Android view and a full-screen version that hides default controls. It includes a SystemUiHider for full-screen mode and a basic layout template.

3. MasterDetailFlow—this template can provide an adaptive layout for your software that can provide different experiences for phone and tablet interfaces. Alternative resource XML resource files are provided for the separate UIs, as well as Fragment, FragmentActivity, and ListFragment implementations.

Activity templates come in five variations. They can be added to any application template by navigating to "New > Other," then to "Android > Android Activity" and clicking "Next."

1. LoginActivity—this is the recommended way to add login capability to your application. Includes the AsyncTask implementation and progress indicator.

2. SettingsActivity—the settings menu will be one of the most used screens in your software. Includes PreferenceActivity implementation and sample values in an XML file to make settings easy to configure.

3. BlankActivity/FullScreenActivity/MasterDetailFlow—the last three activities are versions of the application templates so that you can include the same functionality flexibility with other templates.

Android object templates can be added the same way as activities. The wizard includes instructions for proper implementation of the four following options:

1. BroadcastReceiver—this is a short receiver that allows you to register for system or application events. It also allows the device to be notified once the event begins.

2.   ContentProvider—this allows your software to interact with other applications on the user's device. The more you can integrate with other services, the more value you can provide users from your solution.

3.   Custom View—a template that comes in handy for custom views you design.

4.   Service—not to be confused with a separate process or thread, a service allows the app to perform long-running or background processes.

KEEP GETTING TO KNOW YOUR USERS

Much is currently being made of mobile users and who they are becoming. Terabytes of data are devoted to the coming generation, the types of devices they use, and the activities they are performing on them. It was the same with the television before: I used it for vastly different things than my parents did.

I wanted to pass along some data about current users and what we have learned about them so far. This comes from a spring 2013 survey on mobile marketing data:

- 70% of tablet use comes in the home. I would say that this makes it equivalent to a laptop that is connected to LTE. Certainly places the device in a different realm from the phone.

- Nearly a third of the time spent on a tablet by people ages 18–49 is for gaming. Wow.

- The top features mobile shoppers appreciate are side-by-side comparison, easy checkout, product information, and customer reviews. This means: make the process simple and informative.

- 61% of users want a "click to call" feature on mobile business sites. Most interesting part of this stat is that the word "click" is used in relation to mobile.

- One out of three mobile apps has a "check in" service. On top of that, 48% of users get some sort of incentive for checking in on their device. Once indoor GPS becomes mainstream, you can expect this stat to increase.

Know your users, what they want to use their device for in your stores, and how to get them there. That's the start of a great mobile strategy.

Debugging and Testing

The beauty of building your application in Eclipse is that while you are making changes to your project, the build is running and assigning a default debug key to your .apk file. This helps ensure your code structure is consistent and as free of syntax bugs as possible. To view the application in action, however, you must run a manual build in release mode.

Provided you set up your physical device and AVD properly, you can start the build by selecting "Run > Run" or "Run > Debug" in Eclipse. This will begin the process of compiling the project, creating a default run configuration (unless you have already done so), and running the application in the emulator (or device). To build for both types of devices, you will need to utilize the Deployment Target setting in the run configuration. Eclipse usually detects whether a registered device is connected and makes that assumption for you, but there are times when you may want to have both running at the same time.

This phase is where you see the power of the provided development tools from Google. Eclipse has a Java Debug Wire Protocol (JDWP) installed already, but you can use your own if you so desire. The protocol helps identify issues in communication from the front end to the back. It is part of the Java Platform Debugging Architecture, which uses a library of APIs to analyze the Java framework during the testing stage.

Figure 8.8: Don't forget to test on actual devices. There's always something that can be missed.

You can also utilize a test project (covered earlier) to install a suite of tests using the JUnit framework. To create one:

a.   Select "File > New > Other" to start the setup wizard.

b.   Select "Android > Android Test Project."

c.   Name your project.

d.   Define a path.

e.   Select the existing Android project you wish to test.

f.   Select the SDK you wish to use.

g.   Click "Finish."

You can then populate the project with a test package. Be as thorough as you wish, and don't be afraid to create multiple versions with varying levels of complexity. If you merely want to test the login activity for functionality or performance, make sure to isolate it. Full end-to-end flows are where you can see the true power of the project come to life because it lives and breathes on its own from the existing Android project.

To run the test project, select "Run As . . . > JUnit Test" in Eclipse. It will need its own run configuration just as your Android project does, but once it's set you can iterate quickly. The configuration also helps if you want to run multiple test sets at once to further automate this process. Running from the configuration is done by selecting "Run > Run Configuration"; then follow the instructions in the dialogue box. You can select one or multiple test sets this way.

For a more detailed description of the testing fundamentals Google provides with its SDK, visit developer.android.com/tools/testing/testing_android.html.

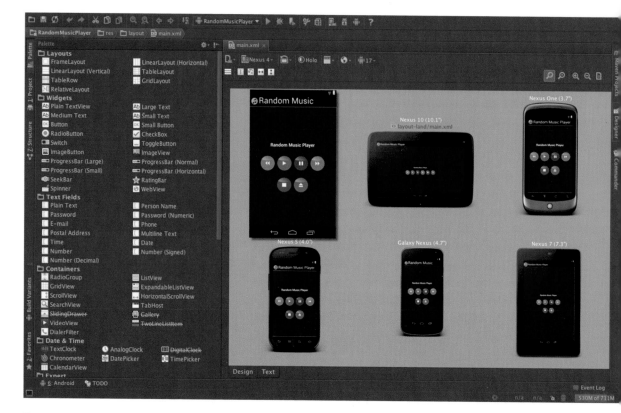

Figure 8.9: Android Studio is Google's attempt to provide a development experience similar to that of Apple's Xcode.

Development with Android Studio

Android Studio is a new IDE based on IntelliJ IDEA, a third-party commercial tool that incorporates several tools and frameworks into one package for developers. Similar to Eclipse with the ADT Plugin, Android Studio provides integrated developer tools for coding and debugging. One could argue that this is Google's answer to Xcode for Apple development, but, just as it's good to have Google's version of an Android device, it's a great idea for the company to provide guidelines for how it thinks software should be built on its OS. Some of the features and advantages Android Studio offers are:

- Syntax highlighting and auto-complete functionality for rapid development
- A WYSIWYG editor for simple editing of graphic files that also allows engineers the ability to preview work
- A plugin for Gradle-based build support
- Lint tools to catch performance, usability, version compatibility, and other problems
- Easier access to Google services from the development environment
- Template-based wizards to create common Android designs and components

One thing to consider: Android Studio is currently available as an early-access preview only. Several features are either incomplete or not yet implemented, and you may encounter bugs. Consult developer.android.com/sdk/installing/studio.html for more information on the current issues and fixes before beginning.

Migrating to AS from Eclipse is an easy process if you are willing to try out the tool while it is in preview mode. Before you begin, you must have version 22 or higher of the Eclipse ADT plugin. In Eclipse, select "File > Export"; then select "Generate Gradle Files." Then select the project you wish to export to Android Studio and click "Finish."

Publishing

Unlike Apple, Android makes it simple to distribute your application through the Google Play store or direct download from your website. There are a few preparatory items needed before you can release your app in the wild for feedback.

- Turn off debugging, unless you want to do a small beta release for feedback from targeted users. The logs generated will only hurt performance.

- You must build and sign a release version, which is guided by the SDK once you are ready.

- Even if you did a thorough job of testing the code during development, make sure you test the release version.

- Ensure that your application resources are updated. Developers often add a device late in the game but forget to create and assign the necessary resources.

- If your app needs a remote server to access backend services, make sure you have them spun up and configured before release.

This is the list of prerequisites, but there might be other items necessary for publication of your software. Make sure you assign a private key or key for other services you have incorporated, such as Google Maps API. Once all of this is in place in a release build, the .apk file you created will be all you need to publish.

One last consideration before publishing is the size of your app. Currently, Google does not allow a file larger than 50 MB to be published. As a rule of thumb, however, you want to be careful about how much space you are asking users to devote to your software. Some OEMs make space difficult on Android devices because of bloatware, as we are all aware. If you do need

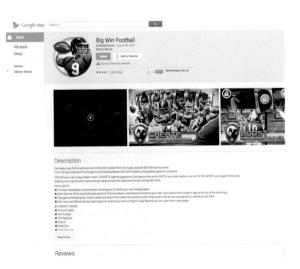

Figure 8.10: More freedom in your app listing means that you must pay even more attention to marketing it.

the space, Google Play does allow APK expansion files to be uploaded and stored separately for developers that need the extra space.

To assist in any needed consolation, Android provides access to a tool named ProGuard to optimize your code base. It analyzes for unused code and renames methods, fields, and classes with more semantically obscure names. While this may have been part of the code review process, it can never hurt to take advantage of a tool such as this. As always, make sure you retest the resulting .apk file.

EDUCATION MUST NOT BE ASSUMED

I once worked for a company that provides Web services to an industry predominantly behind the times technologically: car dealerships. When I first walked in the door, I threw around a lot of terms we used in my previous job, which was mobile focused. You can imagine the looks I got from some (not all) of the stakeholders. I realized that before we could craft a strategy, we had to make sure everyone was one the same page.

The mistake would be to swing the pendulum too far and overeducate. We are talking strategy only at this point. Feel free to be as detailed as you want when you move to an ideation session with product and development teams. As strange as it sounds, focus on a glossary and gaps analysis in your current product offering. The answer will practically jump off the page after that exercise.

Google Play Distribution

While publishing your app on Google Play is a snap, there are still options that need to be determined before you upload and distribute. When you head to play.google.com/apps/publish, simply click the "Add new application" button to begin the process. After entering a title, description, promo text, and recent changes (if you are uploading an update), you need to upload all graphic assets.

Figure 8.11: Distribute, market, and collaborate online with Google Play.

The beauty of Google Play is that you have more flexibility in how you market your product. While screenshots and a high-res icon are required, you are also allowed to include a feature graphic and videos to show your app in action.

After that, there is a short checklist of items you need to determine in the listing. Here is a list of things to consider:

1. Determine the countries in which you plan on distributing your app. If pricing is part of your roadmap, make sure you consider what you are going to charge in the local currencies.

2. There are also localization considerations, which can be as complicated as you make them. Customizations with languages, alternative layouts, and time zones need to be done in the code and assets.

3. Decide on the content rating. There are four settings: available to everyone and low, medium, and high maturity. The more restrictive your content, the less an audience will be able to locate it in the store.

4. You will also need to confirm which device and screen sizes will be compatible with your app. Since you already took this into consideration during the planning stage, this should be an easy choice for you.

5. Pricing does need to be determined at some point before publication. One distinction Google makes is this: if the app is priced as free, it must remain free. Make sure revenue models are in place if you choose this option. Same goes for in-app billing.

6. There are product details you can include in your product listing to help users find your software. Do your research on what category type you wish to use.

Unlike the iOS App Store, Google Play allows you to publish beta versions of your product. Just make sure your users are aware of this and have a clear path to contact support during testing.

E-mail Distribution

Depending on your product category and connectivity with a base of users, it might make more sense to just distribute your product via e-mail. Because all you need is an .apk file, you merely attached it and hit "Send."

Where this becomes a huge advantage is during the testing process. Instead of having to deal with a lengthy device registration process like Apple's, you can just send out your app for immediate feedback. When recipients open the e-mail on an Android-powered device, the device will recognize the file and display an "Install Now" button in the message. Touching the button starts the process.

One recommendation I might have is that while this method may help during development, there is no way you can charge for the app if you use this method. It is also very difficult to protect yourself from piracy and unauthorized distribution, because users merely need forward the mail send out the app without permission.

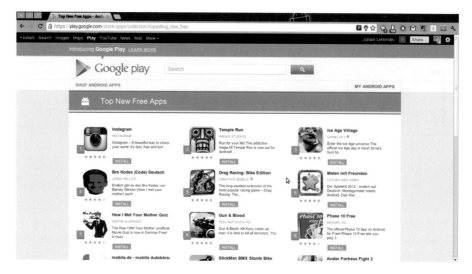

Figure 8.12: Pay attention to the performance of your releases. Your users will tell you what works and what can be improved.

Website Distribution

Private marketplaces are starting to be used more and more by developers. They can be advantageous for companies wanting to distribute internal mobile apps or independents that wish to keep all of the revenue generated. Your app is distributed in the same method as e-mail, except you host the .apk file on your site's server instead of attaching it to a message. Android-powered devices will recognize the file in the link and immediately begin installation, provided the device is set to allow installation of apps from unknown sources.

While you can generate more revenue from your app this way, there are some drawbacks to this distribution model.

- The Google Play store has built in search engine optimization (SEO) and search capability to help users find your software, so you will need to make sure your site has as much marketing behind it as possible.

- You also will not be able to take advantage of the in-app billing service to generate secondary revenue from downloads.

- You also will not be able to restrict which devices are designed to work well with your code.

- Finally, some network providers don't allow users to install applications from unknown sources.

C H A P T E R   9

The Dark Horse— Windows 8

Once upon a time, a company called Microsoft dominated the software industry. Today, Microsoft is the distant third platform solution. Windows 8 is Microsoft's opportunity to turn the boat around and become an equal competitor to Apple and Google.

The focus of this chapter is Microsoft's efforts related to tablet and phone. With that said, you can not talk about Windows Phone and tablet without talking a little about Windows 8 and 8.1.

Figure 9.1: Windows 8 and the new "touch" focus on interaction.

The challenge Microsoft is now facing is exceptional. It is squarely in third place for mobility and losing ground as the OS for PCs. The reason for this is very easy to understand: phones and tablets really can do a lot of the tasks we traditionally complete on desktops. Checking e-mail, accepting invites to meetings, updating social media, and editing photographs can be easily accomplished on a phone. Tablets are now the new mainstay of every meeting. Presentations, documents, and many productivity tools work great on tablets. The desktop is being left behind.

Figure 9.2: The new Microsoft logo reflects the company's new "flat" design.

This does not mean that there is no space for a desktop. There clearly is and will be for decades to come. What has happened over the past couple of years is that consumers are voting with their wallets and choosing to buy different types of computers: computers we put in our pockets, grab like a book, or wear on our wrists.

Microsoft has come late to the mobile smartphone revolution. This is strange, because Microsoft has had a mobile set of solutions since the late 1990s. The question is this: has Microsoft done enough to deliver a clear choice to consumers? Is it a dark horse that needs to be watched?

Windows 8—Phone or Tablet?

Apple has made its digital OS strategy very easy to follow: on the desktop you have OS X, and on mobile you have iOS. Both operating systems have a shared heritage but clearly different functionality.

Figure 9.3: Windows CE was an earlier attempt by Microsoft to bring Windows to mobile devices.

Microsoft's legacy approach to mobility has been this: let's bring Windows to mobile. Earlier versions of Windows' mobile OS, also known as Windows CE (Compact Edition), showed that Microsoft wanted to keep the desktop concepts the same for all devices. To allow you to navigate menus on a phone, Microsoft gave you a stylus.

But it did not work. Both Apple and then Google with Android demonstrated that the mobile OS is not the same as the desktop's.

So Microsoft is now taking a different track, a strategy that goes beyond the PC.

MICROSOFT'S POST-PC STRATEGY

Windows Phone 7 was the first clear demonstration of a new direction for Microsoft. The interface leverages a modern design style known as "flat." Initially, Microsoft called the Windows Phone 7 design style "Metro," but the company was told not to use the name by its legal department. But the name has stuck.

The "metro" style calls for tiles of information coupled with the artistic use of fonts to present data. I have to admit that I really like it. Windows Phone 7.5/7.8 and Windows Phone 8 extended the concept with live tiles. In other words, app icons can now be intelligent.

Microsoft did not stop at phones with the "metro" concept. Xbox 360 users also received the new design treatment. In addition to the new design, Xbox users also have the advantage of using Kinect. Now you can wave you hands, like an extra from *Minority Report,* to skip screens and tap on tiles.

The next logical step with the tiles concept was to bring it to the mother ship: Windows. Windows 8 is designed for tablets. When I first tried Windows 8 on a handheld tablet in September 2011, I was simply blown away. This is it, I thought. Apple needs to pack up its bags.

Figure 9.4: Windows Phone 7 with a flat "Metro" style.

The icing on the cake came when Microsoft said it would start building its own tablets, called Surface—sleek slates that run Windows 8, an OS obviously designed for touch. So what went wrong?

Figure 9.5: All of Microsoft's products are now supporting a tile metaphor for interaction, including Xbox.

DESKTOP VERSUS MOBILE

Microsoft made a bet that it needed to go all in on touch. Great. Awesome. But what about all those billions of computers that run with a mouse, keyboard, and nontouch screen? Yep, you know, like every desktop and laptop running Windows 7.

Turns out that Windows 8 is not great outside the touch universe. Indeed, the biggest problem with Windows 8 is that it was desperately trying to forget its heritage. To confuse matters even more,

Figure 9.6: Microsoft's Windows Surface RT.

Windows brought out two versions of its Surface tablets, one that has the full version and one that has a crippled, low-power version, known as Windows RT.

If you had the money to buy a $600 tablet, would you buy the full-featured iPad or Android tablet (such as the Note 8 from Samsung), or would you buy a crippled version of Windows? Again, consumers spoke with their wallets. By mid-2013, Microsoft had to publicly admit to writing off $900 million on Surface RT tablets that no one wanted.

Hindsight is a great tutor. It is clear that Microsoft overestimated the demand for touch devices with the Surface. It saw touch as a feature for every device instead of addressing the contextual need of each type of device.

The good news is that Microsoft is a big company and can learn from its mistakes. And Microsoft does. The foundation for Windows is great, with some amazing technologies. Microsoft will adapt as Windows 8.1, 8.2, and so on are released.

CODE IS CODE

It is clear that Microsoft is driving all of its platforms to one code base. This makes sense. After all, Apple has been doing the same for years. Xcode for Mac apps is the same as Xcode for iOS.

Expanding the scope of what your code can do has another benefit: it allows you to expand the skill sets of the developers already using your technology. Today there are millions of C# developers comfortable working with Microsoft's development tools. Why not bring these developers along to the new mobile and surface platforms?

This is exactly what Microsoft intends to do. Developers are a powerful army. The changing tide for mobile technologies clearly shows that consumers are fickle. If your development team knows only Microsoft technologies, why not invest in Windows Phone, Surface, and desktop? You have the talent to build something that you can use.

The Tools You Need

Like Apple, Microsoft has a mature set of tools you can leverage to build out the solutions you will need to build apps. Indeed, Microsoft arguably has the strongest development tools. Visual Studio, your development tool suite, is currently on version 2013 and has roots that go back to 1995, when the first version of VS was released.

Unfortunately, Microsoft has not really made it too easy to develop across the Windows platform. You will

Figure 9.7: Windows runs across multiple platforms.

need a collection of tools to be truly successful on the Windows 8 platform. Have the following close at hand:

- Windows Phone 7.5—many of the low-entry Nokia phones (such as 520) are running 7.5, which has different design specifications from Windows Phone 8.
- Windows Phone 8—leading-edge Windows phones are now running Windows Phone 8.
- Windows 8 RT—by the time you read this, you should be able to pick up a Surface RT for just a couple of bucks. Invest. The RT version of Windows 8 is a low-energy version of Windows 8 and has many differences from the full-size edition.
- Windows 8 Tablet—you will want to test your apps on a full version of Windows 8 in "touch" mode on a tablet (the user experience is different from that with a traditional desktop).
- Windows 7/8 development laptop.
- Internet Explorer 11—many companies are banking on Web apps as a future solution, and Microsoft is investing heavily in the Web.

Each environment for Windows solutions is different. Consistency would be a boon if Microsoft applied this idea to its development environment. It is clear with Windows Phone 8 that Microsoft is looking to drive a consistent developer experience, but it is coming late to the party.

It was clear from the first release of Windows 8 that Microsoft was going in a very different direction with the OS—a radically different direction. Sales are showing that the change in direction from Windows 7 to Windows 8 is just too great for consumers.

Enter Windows 8.1 and a commitment from Microsoft to update Windows annually with incremental releases, an OS release cycle similar to Apple's OS X release cycle. The new Windows 8.1 has the same tile metaphor introduced with Windows 8 but has added back fast access to the many features we loved about Windows 7. In many ways, Windows 8.1 is the version of Windows that Microsoft should have shipped.

The tablet interface for Windows 8 is very engaging. There are several key metaphors that make tablet design a compelling direction for your solutions over traditional desktop. They are:

- Full-screen, immersive environment.
- Rapid access to system tools—a swipe to the right of the screen on Windows 8 exposes "Gems," widgets that give you access to file sharing, printing, and other system processes.
- Dual-screen mode—while none of us multitask well, there are times when our task is to edit in one document content drawn from another (such as sending a summary of activity derived from a spreadsheet in an e-mail). Windows 8 has elegantly introduced a split-screen tool for use when two apps are running at one time.

- Windows 8 has Windows running under the hood—yes, the tile metaphor is new, but many of the utilities you are familiar with from Windows are running under the hood.

For these reasons and many more, it is worth keeping a close eye on Microsoft. This is the company that does have Windows on more than a million PCs, and you can bet your last dollar that it will not let go of its dominance easily.

Introduction to Visual Studio 2013

Visual Studio is the center of your development universe. There are many versions of Studio, and this can be confusing. To keep life simple, focus on Visual Studio Express.

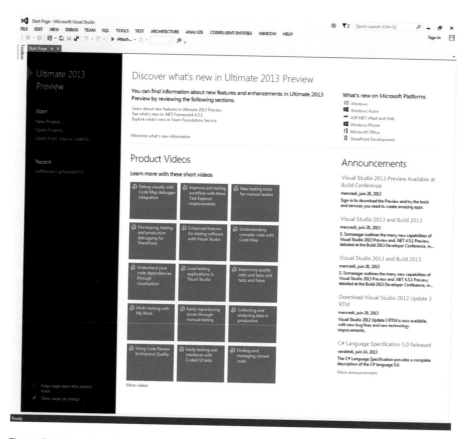

Figure 9.8: Visual Studio.

RAPIDLY CREATING APPS

Visual Studio has always been about giving you the power to rapidly create solutions. An extensive set of tools gives you the ability to drag and drop controls to visually create an application. Need a checkbox? Just drag a checkbox from the control panel and drop it on the screen.

There are millions of Visual Studio developers. To support the massive community, a growing number of companies are selling features that Microsoft may have missed. Need a widget that creates an image library on the fly? Someone has built the feature for Visual Studio.

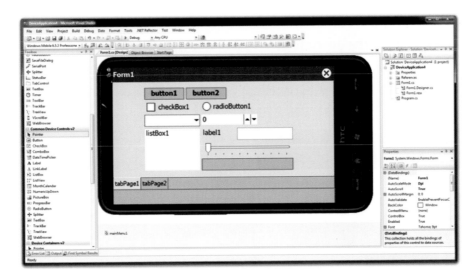

Figure 9.9: Creating solutions with Visual Studio.

Visual Studio is scalable for the amateur developer up to large teams. If you are new to Visual Studio, you will want to start with the Express tools. The Express tools are free to use and come in flavors to create Web, phone, and desktop solutions.

There are a lot of different versions of Visual Studio. Microsoft created this chart to help you decide which version you need: www.microsoft.com/visualstudio/eng/products/compare.

DECIDING WHICH CODE PLATFORM TO CHOOSE—C#, VB.NET, OR HTML

Visual Studio is a development environment. You can choose among three different programming languages to develop your solutions:

- C#
- VB.NET
- HTML

For many developers, C# is becoming the de facto development language. The roots to C# are similar to those of Java and C++. This makes transitioning from one language to another much easier.

Visual Basic, VB, is the development language Microsoft developed in the 1990s. The structure of VB is very different from that of almost any language I know. The goal with VB was to make it easier for you to build solutions. The introduction of the .NET framework in 2001 brought with it a massive overhaul of VB. VB.NET is still a popular language. If, however, you are new to Visual Studio, I would encourage you to learn C#, as it will make your transition from one development platform to another easier.

The release of Visual Studio 2012 brought HTML into the fold as a top-tier development language alongside C# and VB.NET. All you need to do is bring along your JavaScript, CSS, and HTML5 skills to build out a solution.

Working with Windows Phone App Studio

In contrast to Visual Studio is a new product from Microsoft that targets nondevelopers. Windows Phone App Studio gives anyone comfortable working on a computer the ability to create a Windows Phone app through a drag-and-drop interface.

Figure 9.10: Rapidly develop Windows Phone apps from templates in App Studio.

On the whole, Windows Phone App Studio is a template-driven solution, and this is OK for many apps. If you are a small business, nonprofit, or community group or you want only to test the Windows Phone App world, App Studio is an easy tool to test out. Reviews in the press are mixed, but I think that the press is missing out. Many companies can't afford large development groups to create iOS, Android, and Windows Phone apps. Microsoft is doing us a great favor by introducing a tool everyone can use. And remember, this is only version 1.0.

Internet Explorer for Web Apps

The Web browser Internet Explorer has been much maligned over the years. There is a good reason for this: in 1998 Netscape was effectively killed as a company when Microsoft gave away its Web browser. And then came Internet Explorer 6—a crippled browser with no competition and no updates for years.

Figure 9.11: Internet Explorer 11 comes with solid support for HTML5.

And then came HTML5, supported by Firefox, Safari, and Google Chrome. A shakeup at the Microsoft Internet Explorer camp had to happen. The result is Internet Explorer versions 9, 10, and now 11, each reaching to fully embrace HTML5 and Web standards. The result is a great browser. Interestingly, the engine that powers the Web browser on the desktop is the same version that runs on phones and tablets.

Microsoft in competition for the post-native app world of Web apps. For today, native apps rule the roost. However, Google, Samsung, and Microsoft are investing heavily in the Web, and Internet Explorer may be a foundation tool you need to use for future solutions.

Beyond the Dark Horse

It is clear that Microsoft is in a tough spot today. The fact that Microsoft is hardly mentioned in this book is a leading indicator, reflected in the tepid sales of Surface, Windows Phone, and Windows 8.

With that said, Microsoft is a massive company that can keep working at a technology until it is right. Today, Microsoft has the right tools, a solid technology platform, and the opportunity to deliver. Microsoft could well be the #3 player in the mobile world behind Google and Apple. In other words: a dark horse.

But the race is not over. Unlike in the 1990s, there are many companies looking to be in the mobility race. Free solutions such as Ubuntu, Firefox OS, and Tizen all look appealing when you consider that Microsoft still charges for its mobile OS to carriers. Only time will tell who will win.

3

Marketing Your Apps

Publishing to App Stores

Developing an app is easy. Knowing where to publish your app is not quite so easy. The focus of this chapter is to explain how you can publish your apps to the leading stores and understand their review.

Apple's iTunes App Store and Google Play are the two stores I will focus on in this chapter. Combined, these two stores service more than 1.5 billion devices. Today, Google is registering more than 1.5 million new devices every single day; it has already registered a staggering 1 billion devices. For its part, Apple has registered more than 600 million devices. You can put together all of the other device manufacturers on the market and the total will not even come close to Google and Apple.

There are, however, hundreds of app stores outside the two big players. Later in the chapter you will find out about the leading contenders and decide for which companies you should develop your apps.

Finally, the app store that will appeal to many companies is the Enterprise App Store. Enterprise App Stores provide companies with the ability to create their own applications that can be shared privately with employees and strategic partners. In many ways, the Enterprise App Store has the greatest potential for growth over the next decade.

Figure 10.1: The original iTunes App Store launched in 2008.

50 billion app downloads.
One really big thank you.

50,000,000,000

The grand prize winner will be announced soon. Stay tuned.

Official Rules

Figure 10.2: Apple reached its 50 billionth app download in 5 years.

The goal for this chapter is simple: know where and why you are publishing your apps. Unlike the Web, app stores are highly curated private stores. Just because you build it does not mean your app will be published.

Which Store Do You Publish To?

Imagine you are developing your first app. You have this brilliant idea and can't wait to publish. As you have seen in the earlier chapters in this book, writing an app is different for each mobile platform. Each has its own design guidelines and programming languages. Nothing is really the same.

But, maybe, instead of thinking which platform you want to publish to first, you should consider which app store you are publishing to.

The clear leaders are Google Play and Apple's iTunes App Store. Don't even consider any of the other app stores for your first app. These two app stores provide you with the template you need when publishing to all of the stores.

When Google first launched Google Play (then called Android Market), the store felt very much like an experiment. There were few rules, and posting apps was easy. The downside was that your apps were limited to a few app store markets. The new Google Play is mature and comes with tools you need as a business to thrive in the app-economy.

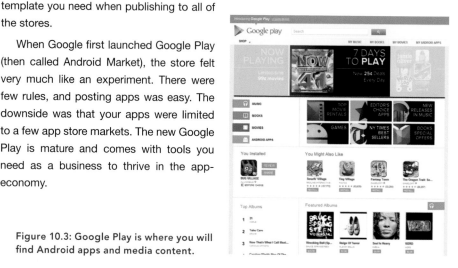

Figure 10.3: Google Play is where you will find Android apps and media content.

In contrast, Apple's iTunes App Store has always felt like a business tool. The concept was simple: treat each app development company as if it is a business. The challenge with Apple is its review process. On one hand, the review process does mitigate duplication of apps (the app submission guidelines state that you do not need to create any more "fart apps," as iTunes has enough) and apps that are malicious. But it is slow! Assume it will take two to three weeks for each app to be reviewed.

So, which app store do you start with? The easy answer is this: which phone does your client have? If the client has an Android phone, then go with Google Play. If the client uses an iPhone, then shoot for the iTunes App Store.

The Value of Icons

There are common themes in developing and submitting apps to any app store. One is the development of an icon. No matter what phone or tablet you have, every app is launched via a tap on an icon.

The challenge is that not all icons are created equally. You must now design your icons in different sizes to meet the different screen resolutions for different mobile OSs. The three main groups of icons are:

- Android
- iPhone
- iPad

Figure 10.4: Icons are the new method of branding your business.

PNG File:
128x128 pixel

PNG File:
96x96 pixel

PNG File:
72x72 pixel

PNG File:
64x64 pixel

Figure 10.5: Different devices need different icon sizes.

For each platform you will also likely create icons that support standard and Retina displays. The result is that you need up to a dozen or more icons for each app you create. The good news is that there are tools you can use to make your life easier. Two of my favorites are:

- iDeveloper Icon Generator
- Empoc Icons

Both tools are available on the Mac App Store. To get started, you will want to create a large 1024 × 1024 pixel version of your icon.

Figure 10.6: iDeveloper Icon Generator.

Both iDeveloper Icon Generator and Empoc Icons require you to drag your image onto the software. From there you choose which platforms you want your icons to be published to.

Figure 10.7: Create icons with Empoc Icons.

Select the Generate or Batch button, and watch all of the icons you need for every platform. Both of these tools make creating icons very easy.

So, now the hard part: your icon must stand out. Both of the leading app stores have more than a million apps. The leading apps for any store have invested time in creating beautiful icons. Spend the same time. Do not settle for a bland icon. Your icon is your book cover, and people really do judge a book by its cover.

Apple App Store

If you have a choice about which app store you should submit your app to, you will want to choose Apple's iTunes App Store. There are number of reasons why:

- More than 93 percent of all iTunes apps are downloaded every month (in other words, you will have an audience).
- The iTunes app store is one of the most complicated app stores to get published to—master the iTunes App Store and most other stores will look easy.
- Apple is picky—be prepared to submit your best work or expect a rejection letter.
- Apple has done a great job of ensuring that the latest version of iOS is running on all devices, reducing the number of devices you need to test on.

Publishing an app to the iTunes App Store is a rewarding experience. It is also a minefield. It took me a full day to submit my first app to the iTunes App Store. Now I have it down to about one or two hours. It is not a fast process.

WHY DEVELOPERS SELL MORE ON THE iPHONE

With all of the challenges Apple throws at you, you may wonder why developers often choose Apple over Google Play. After all, Google Play is really easy to submit apps to, and nearly 80 percent of all smartphones run Android.

There are two problems with Android. The first is fragmentation. Android is an open-source project and can be modified by anyone. You can create your own custom version of Android. To this end, we see some wild versions of Android, such as the version Amazon uses for the Kindle. Even today, I find that a third of all my apps are being run on Android phones running a version of the OS that came out in 2010.

The second reason is a personal observation: Android users do not seem to download apps as much as Apple device owners. No idea why. If you have a hunch for why this is, please send me a note on Twitter to @matthewadavid.

With both of these statements said, it is clear that Google is working hard to address both problems, and, in many ways, the new tools Google is now realizing are leaving Apple in the dust. Google Play, the reimaging of the original Android Market, is much easier to use. The latest release of Android also self-updates if the carrier allows the device to do so.

Today, you will make more money from Apple, but keep a close eye on Google. At the end of the day, as developers, we win.

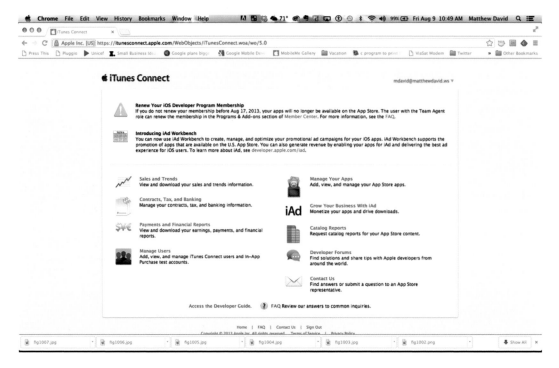

Figure 10.8: Manage your iOS and Mac apps with iTunes Connect.

BUSINESS TOOLS FOR DEVELOPERS

Apple's iTunes App Store is the front-facing storefront to buy your app. The business side is known as iTunes Connect (itunesconnect.apple.com).

The focus of iTunes Connect is to give you the business tools to run your app business. The home page for iTunes Connect is broken into the following sections:

- News and Alerts—across the top are news and alerts, along with a contracts/legal documents you need to sign.
- Sales and Trends—watch how much money you are making from your apps.
- Contracts, Tax, and Banking—you will need to spend the time to set up your bank and tax information with Apple to ensure you get paid correctly.
- Payments and Financial Reports—detail by global region for all of the money you will make.
- Manage Users—as your business grows, you will need to provide access to sections of iTunes Connect to different people (e.g., developers, accountants, sales managers).
- Manage Your Apps—the heart of iTunes Connect is managing all of your apps.
- Growing Your Business with iAds—tools to add Apple's proprietary in-app advertising program.
- Catalog Reports—detailed reports on your apps.

- Developer Forums—got a question? Ask the forums. They are awesome.
- Contact Us—have a question regarding an app submission? You can contact Apple directly.

Apple is always adding and improving the iTunes Connect experience. Treat Connect as your strategic business partner, no matter how large your business is. For many years I was a development company of one person. Today I have 300 people on my team. We use the same iTunes Connect.

CONTROL OF APP RELEASE

To be clear, I am not going to step through the details of submitting apps to iTunes Connect. There are many great books you can get that dig deep into the details. Also, the best exercise is actually to submit an app. The best piece of advice I can give you when you submit your first app is to give yourself lots of time the first time to get through the process. A full day would be good.

Figure 10.9: Manage your apps with iTunes Connect.

The app submission process can be controlled a number of ways. Apps can be submitted directly from Xcode or through the Application Loader. Here is the caveat: you must submit your apps from a Mac. No Windows love here.

WEIRD APPLE RESTRICTIONS

Many clients I work with are still incredulous that I cannot build iPhone apps and submit from a Windows machine. Let me set the record straight: you can't. Don't even try. If you want to get into the iPhone app business then you must invest in an Apple Mac.

Apple does a really great job of letting you know where your app is in the review process. For instance, you will receive e-mails when your app is uploaded successfully to iTunes Connect, when your app is in review, and when you app is being processed for submission to the iTunes App Store.

In addition, you also have access to a dashboard that gives you an update on the status of your apps. This is great when you are managing a lot of apps.

MANAGING THE APP APPROVAL PROCESS

Apple rejects roughly half the apps that are submitted to it, so you should expect a rejection rate of about 50 percent. This has been a consistent trend I have seen since my first app was submitted, in 2009. There are a number of reasons why your apps will be rejected, including the following:

- This is your first app—Apple often rejects the fist app submitted by a new developer. I see this behavior as very similar to taking your driving test—examiners almost always fail you the first time just to make sure you will work harder and come back.

- Your app is too similar to a group of other apps.

- Your app is too similar to a book—if you have a reference app that feels like it could be a book, then Apple will ask you to submit to the iBook Store.

- Your app is a copy of a website.

- Your app is running illegal code—don't try to slip a virus or malware into your app, as Apple does check for this. If you do, you will lose your developer account as well as having your app rejected.

Apple has conveniently put together a set of guidelines you can review to help assess the type of app you are submitting. You can view the guideline here: https://developer.apple.com/appstore/guidelines.html. On the whole, the guidelines are very serious, but there are also some gems Apple has take the time to add, including the following:

- We have over 700,000 Apps in the App Store. If your App doesn't do something useful, unique or provide some form of lasting entertainment, it may not be accepted.

- If your App looks like it was cobbled together in a few days, or you're trying to get your first practice App into the store to impress your friends, please brace yourself for rejection. We have lots of serious developers who don't want their quality Apps to be surrounded by amateur hour.

- We will reject Apps for any content or behavior that we believe is over the line. What line, you ask? Well, as a Supreme Court Justice once said, "I'll know it when I see it." And we think that you will also know it when you cross it.

The good news is that when Apple rejects an app, it will give you guidelines for why the app was rejected. If, however, you feel that the rejection was not justified, you can appeal to the App Review Board at https://developer.apple.com/appstore/contact/?topic=appeal. Yep, good luck with that.

THE IMPORTANCE OF PRICE, REVIEWS, AND FEATURES

For most of us, we want to develop apps so that we can make some money. There are indie developers who have made lots of money (check out the game Tiny Wings, developed by one person, which was in the top ten for nearly a year). You will notice with all popular apps that the development team has taken time to highlight all of the unique features of the apps. There are a million apps in the iTunes App Store—focus on what makes you special.

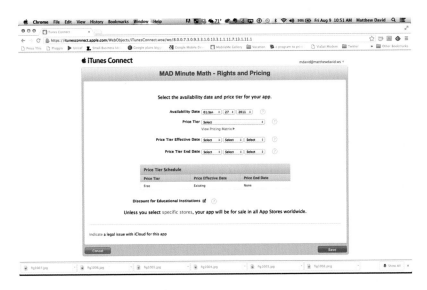

Figure 10.10: Managing app price.

Price is hugely important for your apps. Here are ways to make money with your app:

- Make it free—Zero is a magic number, and your downloads will increase when you give away your hard work. But there is not much money in "free," so check out the next two tools to bring in the loot.

- Ads—advertising is an easy way to generate revenue for your apps.

- In-app purchase—almost every free app now gives you the option to gain access to new content by buying features inside the app.

- Setting a price—Apple leverages a fixed, zone-driven pricing model. For instance, if you price your app at $1.99, it will appear as €1.99 in France. Depending on the exchange rate, you may make more money from French downloads than from those in the United States.

Apple will take a commission of 30 percent for each virtual penny you earn. Bear this in mind when you price your apps.

Interestingly, almost all of the apps in the iTunes App Store are now free. Revenue from ads and in-app purchases often exceeds income from any fixed price.

In order to get an increase in downloads so that you can make more money, you do need to get good reviews. For a while you could actually pay companies to write reviews for you. Unsurprisingly, Apple is now stopping this behavior. When it comes to working with Apple, honesty is the best policy.

RELEASE OFTEN

Once upon a time, development teams had to release an update only every 18 months. Today, the expectation is very different. Release, and release often.

Do not force everything into your app. Get version 1.0 out of the door as quickly as possible. Your customers will tell you what you are doing well and what needs to be improved on.

Build in features over time.

Google Play

In 2012, Google rebranded Android Market as Google Play. It was a brilliant move. With your Google device, you can now access a broad selection of media: apps, music, movies, books, and magazines.

In many ways, Google is losing control over the Android brand. The core reason why companies such as Samsung, Sony, and Amazon are using Android is that the mobile OS is open source and can be changed. There is no consistent look or feel to Android with the exception of on Google's own devices.

To this end, Android is now seen as a foundation technology that is important to developers but has little to no value to consumers. This is why the Android Market is gone and Google Play is now here.

ATTRACTING THE 1.5 MILLION NEW ANDROID DAILY ACTIVATIONS

The number of daily activations for Android devices is staggering. Each and every day more than 1.5 million people activate a new Android device. The trend is only growing. Indeed, by the time you read this book, Google will be likely activating more than 2 million devices a day as it marches toward 1 billion device activations per year.

The challenge with Android is fragmentation. Google is working to rein in the many different versions in use. To gain a perspective on the fragmentation, go into your local Walmart, and, in the phone section, check out the prepaid phones. You can get a smartphone running Android in the prepaid phones. Look at the small print before you buy. The phone is running a version of Android from 2010.

In contrast, 93 percent of all iOS devices are running the latest version of iOS.

Today, I find that I must support Android version 2.3 (Gingerbread). Otherwise, I will miss 25 percent of my market.

The potential for reaching billions of customers through the Android OS is clearly there. You will, however, have to run the gauntlet of fragmentation to reach your market.

A BUSINESS APPROACH TO SELLING ON GOOGLE PLAY

Play now supports many tools similar to Apple's iTunes Connect. This is a blessed relief. You can set up your apps on Google Play to be a business and not just a hobby. A big difference between Apple and Google is that Google lets you choose any price for your apps. This allows you to set your price at $1.00 or $250.45—there are no restrictions as there are with the Apple Store.

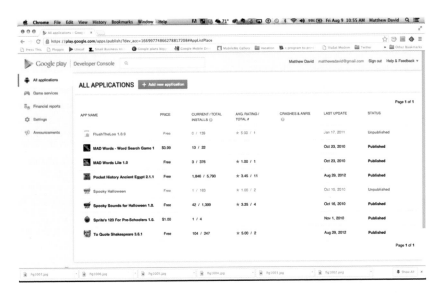

Figure 10.11: Google Developer Console.

A new set of features in Google Play allows you to set your app to run in "Demo" mode for a period of time. This gives customers the opportunity to download your app and take it for a spin. After the "Demo" period expires there is the option to buy the app.

Additionally, you can also release beta versions of your app early. Again, this gives you the opportunity to build customer awareness. When your app goes live, users have the option to buy the full version of the app.

BUILD FOR TABLETS

Google is moving aggressively into the tablet market. Apple with its iPad is dominating that market. The power play Apple often quotes is that it has 300,000+ apps designed specifically for the iPad and iPad mini. Google does not have anywhere close to that number for Android devices.

But this gives you an opportunity. Apps designed for tablets running Android are now getting their own promotional space on Google Play.

While Apple does currently have the lead for tablets, it is clear that its competitors are gunning for them. Google's Nexus 7 is a great tablet (my personal favorite), and the Samsung Note 8 is a huge hit. If you are not building apps for Android tablets, now is a good time to consider a new market.

Other App Stores

Apple's iTunes App Store, released in 2008, was an overnight success. Immediately, hardware manufacturers saw that their devices could be gateways to highly curated areas for consuming content. Content can be commercialized. This is a paradigm that flies in the face of the open Web.

To be clear, the Web is not going to die. Today, devices are providing more ways to connect to the Web than ever before. What we now have, however, is a way to reward businesses that invest in content. Mobility is creating a new business revenue stream.

Outside Apple's iTunes App Store and Google Play, there are quite literally hundreds of app stores. This includes leaders such as the Amazon App Store, Barnes and Noble NOOK App Store, and Microsoft Windows Phone Market, but you also have many niche players. For instance, your Blu-Ray player likely has apps, many TVs now have access to apps, and even your car can now have apps.

There will be more and more app stores to support the ever-increasing number of devices hitting the market. Smartwatches, wearable devices, and intelligent machines will all need their own custom apps to support their niche service.

WORKING WITH AMAZON'S APP STORE

Amazon is the third largest app store. The latest Kindles can now be augmented with apps. Yes, you can read your book and then play a round of Angry Birds.

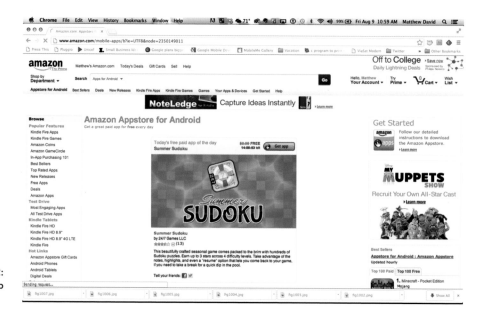

Figure 10.12: Amazon App Store.

Amazon's app acceptance process is very much a halfway house between Apple's strict store submission process and Google's loose philosophy. Some of the caveats include:

- Each app must run on Amazon's version of Android (a branched version of Android 2.3).

- Each app you submit to Amazon is reviewed.

- You can use your existing Amazon account.

Like Google Play, Amazon is rapidly building out its app store experience for customers and developers to meet and exceed Apple's tools.

USING THE BARNES AND NOBLE NOOK APP STORE

A struggling contender in the eBook market is Barnes and Noble. The NOOK is a great tablet, but competition is forcing Barnes and Noble's efforts to the sidelines. Unfortunately, it is too easy to duplicate digital book, magazine, and media content. Physical stores are no longer the only channels through which we can get our content.

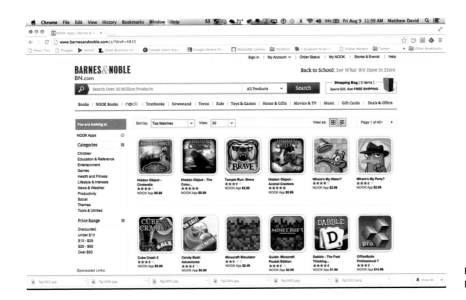

Figure 10.13:
NOOK App Store.

With that said, the NOOK is a great tablet. It is running the latest version of Android, which makes building apps much easier.

There are always winners and losers in technology. Remember Betamax versus VHS? Do not count NOOK out just yet. Barnes and Noble has openly admitted that it wants to spin off NOOK, and Microsoft is a big investor in the device. Round one may be to Amazon, but the eBook fight is far from over.

MICROSOFT MARKETPLACE

Remember Microsoft? It used to be the only way we could connect to the Internet and use technology. The past few years have been hard on the team from Redmond. The Windows Phone is struggling; Windows 8 RT died on release; and Windows 8 is still in single-digit adoption.

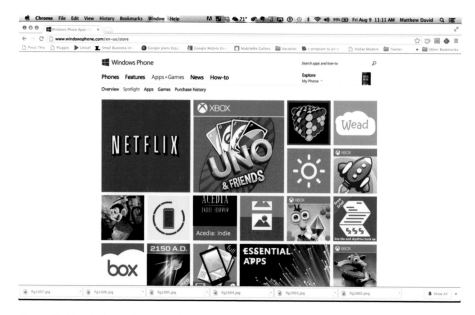

Figure 10.14: Windows Phone Marketplace.

And this baffles me. The new, sleek flat design for Windows Phone and Windows 8 is world class. I really like it. Indeed, the only reason I did not jump from iPhone to Windows Phone is the limited number of apps. Microsoft has really done a great job in providing you all the tools you need to successfully submit and manage your apps through its online stores.

Microsoft is still one of the largest technology companies. It has the time, resources, and money to keep working on Windows until it's right. It is not Apple that needs to worry. It is Google and BlackBerry. Google is offering a free product with the Android mobile OS, but the OS is plagued with malware and virus issues. BlackBerry was the incumbent mobile OS until Apple came along. BlackBerry may still come back. Time will be the judge of this new battle.

The Enterprise App Store

The public app stores get all of the attention. They are cool and shiny and have Angry Birds. They are not, however, practical for companies that want to take advantage of mobile devices in their daily practices. Organizations are now building apps that are being used by employees and partners.

Figure 10.15: Bring the power of mobile apps to your enterprise.

There is a big benefit to developing your own apps. For instance, if your employees are performing tasks that require them to move from one location to another, why not let them use their phone or a tablet to record their work as they perform it? In other words, free your workforce from having to go back and record information on a computer.

In many ways, I see the enterprise market for apps as the fastest-growing area for mobility over the next decade.

BUILDING APPS FOR YOUR OWN COMPANY

The methods outlined in this book for developing applications also apply to enterprise apps. You can choose to build Web apps, hybrids, or native solutions.

The top concern I would encourage you to address when you build enterprise apps is security. The device consuming the data maybe a company-owned device, but it is likely that the device is Bring Your Own (BYOD), that is, owned by the employee. To this end, always assume that every device is not secure. Paranoia is awesome.

MANAGING APP POLICY

Policy enforcement is a necessary evil when it comes to managing devices in your company. There are a number of ways you can easily add policies to devices. They include:

- Exchange Active Sync
- Mobile Device Management (tools such as Airwatch, MobileIron, and Afaria)
- Virtualization

A plus point for mobile device management solutions is that they often come with a solution that allows you to deploy apps. You do not need to create your own enterprise app store.

Try each solution to see which will work best for your company.

A New Store for Every Day Ending in "Y"

The app store landscape is fragmenting. Yes, we have big players who have a massive lead. However, the possibility for a new product to become an overnight success has been documented over and over again in the mobility world. The end result is providing the greatest number of opportunities for your customer to gain access to your product through whichever app store is leading today.

Making Money from Apps

Now that you have published your app, it would be easy to think that the hardest part of your journey is completed. You've scrutinized every user interaction and fine-tuned the performance. Congratulations would seem to be in order, right?

Except now you have to figure out how to make money from it.

Mobile software engineers are in a unique position in today's market. You are not giving users a mobile alternative to a Web service. The desire is to have the native application be the first choice for users. To give users an alternative to current software solutions, you have to do your homework to make your app known to prospective users and price it accordingly.

Selling Apps

Back in its 2010 report "The Future of Application Stores," Forrester, a research company, stated that the discovery and merchandising of mobile applications would be the "next great battleground" for the industry. In more ways than one, that prediction has come true. Even

Figure 11.1: With so many revenue options, make sure you make the right choice.

though the iOS platform has a single place to search for and purchase software, there are nearly a million choices for users. For many developers, the only hope for getting their apps discovered is to hope that they get listed on the main page or top charts. In China alone, there are more than 70 different Android-based app stores. Amazon is adding to the confusion with Kindle apps on both major platforms as well as its own app store for Android.

Needless to say, there are far too many choices to hope that users stumble onto your app on their own. That's why a concise marketing plan is needed. First, let's figure out what to charge. There are several common pricing models on the market today, and each has its pluses and minuses. We discuss the most common in the first part of this chapter.

FREE APPS

According to a survey published by the app analytics company Flurry in July 2013, the average cost of an iPhone app is 19 cents. More than 90 percent of app store offerings are free, and that percentage is expected to keep growing. Many developers have asked themselves how this could be possible. The number is, of course, misleading. Many apps that are free end up creating a ton of revenue through in-app purchases (which we will discuss in a minute). The idea is that many development companies think a lot of software should be free.

Plus: The user can try your app out for free and is always able to use it. This can be helpful if you want to establish a name for yourself or introduce the first in a line of products.

Minus: You make no money unless you have a plan for the future.

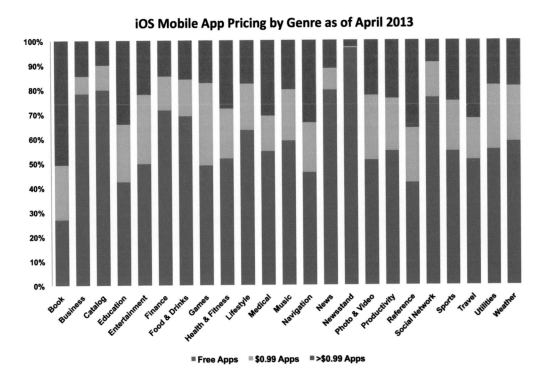

Figure 11.2: The genre of your application will play a big role in determining the pricing model you should choose.

ADVERTISING

In the old days of mobile platforms, iOS was known as the platform where users were willing to pay for software, whereas the Android OS was where you could make money from advertising. As such, apps such as Angry Birds cost 99 cents on iOS and were free (with display ads) on Android. The crazy part was that, as popular as the game was originally on iOS, some touted the revenue on Android as making the app profitable.

Today, much has changed. More Android users are willing to pay according to analytics, and Apple has come a long way in developing advertising models for iOS. With some of the benefits of the iAd Workbench and premium products signing on the dotted line in Cupertino, Apple is doing its very best to help out developers who don't want to charge their customers for software.

Plus: The user can download the software free of charge while still generating revenue for the developer.

Minus: Have you ever had a pop-up ad on your mobile device or have to wait for a video to end to get to the next level of a game? These can be among the most negative, disruptive experiences a mobile app can provide.

REAL-TIME BIDDING IS THE NEXT FRONTIER IN MOBILE ADVERTISING

RTB is not a new concept by any means, but it has made its way to the advertising world. Platforms are being launched that allow advertisers to bid on space in real time on the basis of data analytics and behavior mapping. Instead of having to plan media buys months in advance, you may find that mere minutes of research are all that separate you from potential customers.

There are some challenges to applying this to the mobile platform. Online behavior can't be tracked for mobile audiences in the same way fixed-space Internet media use can. However, it may be only a matter of time before your location, device activity, and data connection can be used to allow companies to deliver relevant advertising to you.

Of course, many will roll their eyes because of the desire to keep personal location and activity data private. All of that is possible by restricting the access certain apps have to that information, but advertising cannot be kept at bay in your mobile devices forever. Increasingly, ads will forever be continually shown to us.

ONE-TIME CHARGE

While most apps are free, the one-time-fee model is the most common for software that does come with a price tag. The developer asks the user to invest a small amount of money, generally 99 cents or $1.99, for a chance to experience the application.

Plus: The user does not have to worry about anything after the upfront cost. The two most common examples of successes with this model are games and mobile access to Web services (such as Instapaper or ESPN radio).

Minus: Apps must be competitively priced unless the user base is already established. This can limit profitability if infrastructure costs rise faster than the user base.

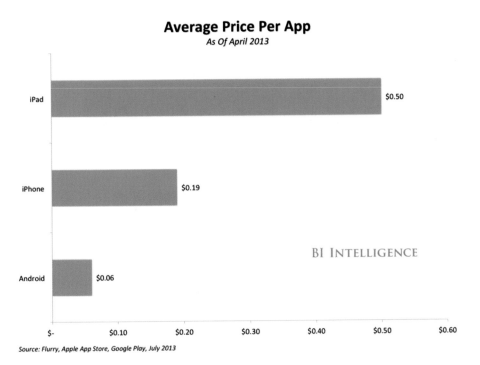

Figure 11.3: While it's not as important as it was in the past, the platform of distribution should be a factor in pricing.

VALUE-BASED PRICING

A version of one-time pricing, this model bases the price of a good or service on the value, perceived or estimated, to the customer, rather than on the cost of the product or historical prices. You should answer two questions to set up a price sensitivity meter:

1. What price would be considered too expensive?

2. What price would be considered too cheap?

After that, a balance between the two is the value-based price.

Plus: Current market trends allow the software to be appropriately priced and stay competitive with other companies.

Minus: It is a rear-facing model that allows companies to only keep up with the pack, not lead it.

USAGE-BASED PRICING

In a usage-based system, users can take part in the full features of the software and be charged only for what they use. Value is generated by a customer's perceived need for the software. This model is most successfully applied to file-storage software such as Box or Dropbox.

Pluses: Users pay a monthly fee for what they currently use and generates more profit from there. Customers' need for the service usually outweighs concerns about usage.

Minus: If your software finds a niche in small-usage customers, the margins start razor thin and only get thinner the more business you acquire.

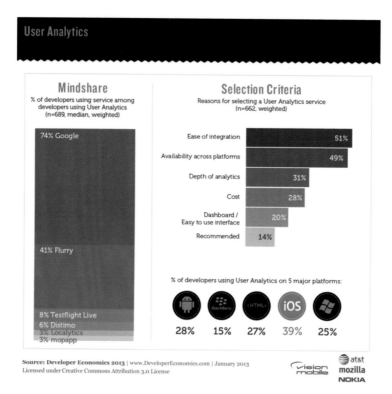

Figure 11.4: If you aren't utilizing analytics, you are missing out on key information to help push your app to the top of the charts.

SUBSCRIPTION

This is the most common form of software licensing in the enterprise sphere. It is used by all the leaders we seek to compete and partner with. Mobile applications are extensions of a service most users can access via the Web front end. The app is free to download, and users can access their account with the help of a sales associate or via the Web.

Plus: Usually, in-app revenue collection is not an issue since the site is the most common form of account management.

Minuses: This model has been perfected by companies that are not mobile-centric. Most users of this type of model are going to use the Web more than native apps.

BEWARE OF MOBILE DEVICE PRIVACY STATS

A study about mobile device privacy released by the nonprofit organization MEF presented a number of alarming statistics about how secure our mobile devices are. In evaluating the numbers, it is important to view this survey of 9,500 respondents in ten countries in the right light. Here are some key takeaways:

- In terms of comfort, 52% of respondents are not at all comfortable storing their credit card information within an app. This is not a huge shock until you take into consideration that payment information is stored on the phone itself. There's a good chance that these respondents used the stored credit card number to purchase many of the apps on their phones.

- In the same vein, 33% are not at all comfortable sharing their personal information, and 35% feel the same way about sharing their location. This means that people are more comfortable sharing where they currently are than their credit card number. Just remember, plastic can be canceled with one phone call, but you can never be unstalked.

- A third of respondents (33%) feel that they have control over how their personal information is used for advertising purposes. This means that one-third of people lie to themselves in the mirror every day.

- With regard to gathering and sharing of information, 70% of respondents feel that it is very important to know what information an app is gathering and subsequently sharing with third-party services. This means that users want to know what Facebook is gathering and then sharing with Spotify. The comment from the preceding paragraph applies here as well.

Don't be afraid to look hard at data when making product and design decisions for your users. While there is still a need for security and awareness, often it is meant to push an agenda irrelevant to your app.

RULE OF 17

This is a subscription-based model that bases the monthly fee on 1/17th of the license fee. A $1,700 piece of software would thus cost $100 a month for two years. Some models run for three years. While this model is considered by some to be outdated, it certainly has lots of validity for enterprise-level software. Companies that provide mobile portals to their large suites of software such as SAP still utilize this model.

Plus: The owner of the software generates extra revenue, charging "financing" fees to space the charges out.

Minus: The user is not obligated to pay for the entire license fee.

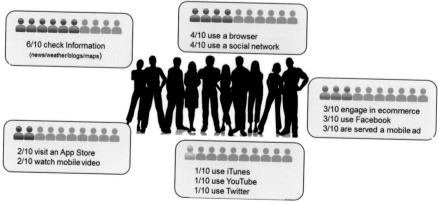

Mobile Subscriber Behavior

On a Typical Day.... Out of 10 Smartphone Subscribers....

6/10 check Information
(news/weather/blogs/maps)

4/10 use a browser
4/10 use a social network

3/10 engage in ecommerce
3/10 use Facebook
3/10 are served a mobile ad

2/10 visit an App Store
2/10 watch mobile video

1/10 use iTunes
1/10 use YouTube
1/10 use Twitter

Figure 11.5: Know your users; they will decide whether or not your app makes money.

FREEMIUM

Unless you are new to mobile apps, you know this is the current standard for consumer app pricing. It is most appealing to users because they get a taste of the user experience and can get as much out of the app as they desire. The app is free with some basic features, but the full user experience will cost the user something via one of two models: upgrades and modules.

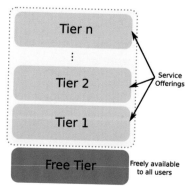

Figure 11.6: Freemium is the current king of app pricing models.

Tier n

Tier 2

Tier 1

Service Offerings

Free Tier

Freely available to all users

Plus: This is the most successful app model in the consumer market right now. The upgrades mirror as "impulse" buys that aid the user in getting through a game faster, or the user may realize that the feature she wants is part of a paid module.

Minus: Almost all of the apps in this model do not require the user to purchase these extra features. It is possible for a use to download and use the application without any compensation to the developer. Some apps incorporate advertising revenue. A freemium upgrade could include the removal of advertisements from the user experience.

WHICH IS THE BEST OPTION?

There is no magic bullet for price. There are successful examples for every pricing model. That said, there is one model that is having the most success in today's market: freemium. In most cases, the free download of your app is the best choice to get users accustomed to your design and user experience. Once the hook has been set, so to speak, the rollout of attractive features allows you to generate your revenue. The key decision is how to package features and sell them at an enticing price point. The subscription plan Google Apps for Business uses is one of the most successful among small business leaders. While Google

Apps for Business is not the only purveyor of these great concepts, it's clear the company follows these practices:

1. It wants you to try before you buy, without having to talk to anyone. This is a huge advantage for companies with mainstream brand recognition because they can leverage their name to let you truly experience how amazing their suite of apps can be to help run your business. Don't be cheap about the free download, either. While you don't need to show all the way behind the curtain, don't think that a lengthy signup process with some fancy animation is the user experience consumers are after.

2. It makes its prices transparent and easy to access. This is something to remember when you are introducing your software to a mobile-device user. Customers have proved they are more likely to accept pricing if you are proactive about disclosing pricing information and what you get for your payment.

    Keep in mind this does not apply across the board. If you want to charge for power-ups in a game, it is best to let users play a few levels before they realize that progressing faster might cost them a little. For an enterprise productivity app, however, you might disclose that a free download allows users to access some great features. If they would like even more, however, a small fee will unlock the door.

Figure 11.7: First taste is always free. Then, using the in-app purchase screen, users will pay if your app works.

3. It gives two prices: monthly and yearly. My uncle Bob used to always say that something we wanted was "just a monthly payment away." In many industry segments, it's still the case that paying a little each month is the best option for consumers. Online purchasers, however, are becoming a little savvier about who they give their credit card info to and what it buys them. If a service costs $10 a month but $100 a year, it makes sense to save yourself $20 and buy a year up front.

4. It makes access to its products easy once you have signed up. Nothing makes users more frustrated than purchasing a service either on a website or in a mobile app and then having a hard time getting to what they want to use. If paying a fee

allows users to get past your pay wall, make sure they don't need to contact support to find the additional content. Also, if users need to get past a security screen to access paid content, make sure logging in is simple.

5.   It provides sales support as more of a "last resort" and not a necessity. We don't like talking to people on the phone unless we have to, right? In the auto industry, car dealerships are now reporting that it is easier for car salesmen to engage their customers through text and e-mail than by phone. That was once heresy in the customer service segment.

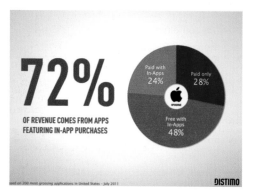

72%
OF REVENUE COMES FROM APPS
FEATURING IN-APP PURCHASES

Paid with In-Apps 24%   Paid only 28%

Free with In-Apps 48%

DISTIMO

Figure 11.8: iOS is riding the freemium model for all it's worth right now.

It's always good business practice to give a phone number for someone to call if there is a problem, but the large platforms don't require that. If need be, an e-mail address is all you need. That said, make sure you answer your e-mails from users quickly and efficiently. If your service becomes so popular that this becomes untenable, that means you must be doing well enough to hire some help. Make sure your support is unobtrusive and easy to figure out.

Depending on the category of your app and the desired consumer base, once the price of your software has been determined, the main decision will be how you plan to collect the revenue. Even though mobile commerce (m-commerce is very quickly replacing e-commerce in mainstream lexicon) is a relatively new field of payment, it has grown by leaps and bounds. Previous roadblocks such as usability, security, and processing fees have been replaced by commerce software like Square and Paypal.

The downside is the fees that companies like Google and Apple pay for collecting payments for you. Both charge 30 percent of the total, which for some can mean a huge difference in profit margins. In the old days, your app would be removed if you even gave users the option to pay for subscriptions or user logins on a separate website. After enough litigation, however, the companies relented. As long as you don't give a link to the website where you can pay anywhere in the app, there is no issue with your keeping the whole fee. The key is to remember to market your website properly. Let consumers know where to go to find out about the company, and let them figure out the rest.

App Discovery

Of course, no matter how appropriate your pricing is with regard to your competition or to business needs, if nobody can find your app, it is all moot. Although the iOS App Store is only five years old, there is a much longer track record of success in optimizing content from the Internet. Search engine optimization and marketing (SEO/SEM) is now a legitimate service that Web-based software companies offer to their customers. It makes sense to apply the same logic to content stores that have grown as vast as Apple's and Google's.

With the size of both stores, however, there is a lot of noise. The days of submitting code and waiting for your name to be featured in the list of "what's hot" on the main page are gone. It's impossible for the two companies to know every single piece of software worth checking out without some sort of assistance.

In the end, you need to make sure copy within your in-store marketing material can point people to your listing. According to 2012 research done by Nielsen and by Forrester, searching within the app store is still the most popular method for users to connect with new software available (at 63 percent in both surveys). Having your app listed in the top charts isn't even as appealing as it once was. In the Forrester survey, the number of respondents who connect using app store top charts is half the number of users who connect via traditional search methods (34 percent).

This finding is only bolstered further by the comments made in 2013 at Google I/O conference by Ankit Jain, head of search and discovery at Google Play: "For the average app, search actually makes up the vast majority of installs."

Figure 11.9: Maximize app store search capabilities with the proper optimization.

KEEP AN EYE OUT FOR PLATFORM CHANGES

In April 2013, Apple announced that it would start rejecting apps that utilize universal digital identifications (UDIDs), which could mean great things for the platform and third-party developers.

While it may seem incongruous to most, allowing apps to track specific devices has actually held back growth of iAd. Apple's platform for delivering advertising revenue took a hit back in 2011 when two high-priority apps (Pandora and Weather Channel) were named in a suit because they did not allow users to opt out of targeted ad tracking. Many, as the article stipulates, have pointed to this as the reason why more big-name brands aren't utilizing the popular platform more.

As of May 1, 2013, the gray area regarding this ruling was cleared up. Currently, iOS is the most compliant digital advertising platform in existence, including PC Web. I can only imagine the positive implications in the next few months.

SEO Your App

So how do you optimize your app for search engines? The major app stores don't all list their apps in the same manner, and even if you follow the instructions to the letter it may not be obvious how you can use the listing to your advantage. For the purpose of this discussion, let's repurpose the phrase "search engine optimization" (SEO) into its app-store equivalent, App Store Optimization (ASO).

The good news is that ASO is in its relative infancy compared to its Internet brethren, meaning you can use tactics similar to those used by Web developers in the late 1990s. Just as you need to build a Web page with the proper metadata, you need to do the same for marketing data. Pay attention to fields such as the app title, description, and keywords. While a copywriter can provide some benefit in this area because she can make it read better, it is not necessary.

App Store Visibility   **Search Visibility**

Figure 11.10: Search visibility feeds app store visibility, which . . .

Your text just needs to parse well without being too "spammy." Giving your app a catchy name will help when people start using it in daily vernacular, but don't be afraid to add a short description of what your app does in the title. An example would be the listing for Strava Run on the iOS App Store: "Strava Run—GPS Running, Training and Cycling Workout Tracker." Notice how the title itself may have helped users to find it, but the subhead allows far more searches to find this software than if the title had stood on its own.

The description field is the one area that doesn't usually impose a limit on your character count. This is where optimizing your copy is most beneficial. If a site reviews your app or a publication lists you among the top new developers in the ecosystem, be sure it is listed above the fold in the description. You may not want to constantly update the description copy, but if you have some bug fixes you want released, make sure this listing is updated with the latest going-on.

Conversely, the keyword field is usually much more restrictive. Apple allows only 100 characters, so it is very important to make the most out this. As with SEO, you must make the copy relevant and easily parsed. Don't use multiple word phrases, repeat yourself, or use spaces.

Also, do your homework, meaning that you should test out the keywords you plan on using. Unless you are inventing a brand-new category of software, it is important to use text similar to that used by apps you want to compete with. If someone is frustrated with an app he has used before, he might search similar keywords and look for an attractive replacement.

Non-Metadata Marketing

Not all of us are designers; a quick browse through any app store can reveal just that. It is important to make graphics such as your logo and screen shots appealing to app store shoppers if you are to have a chance of earning a download. Will your logo serve as a reminder to the user that your software is worth her time?

One of the first books I read years ago on logo design described the graphic as a physical description of your product's usefulness. Please allow me, as one of the authors of this book, to use a personal story to convey this meaning.

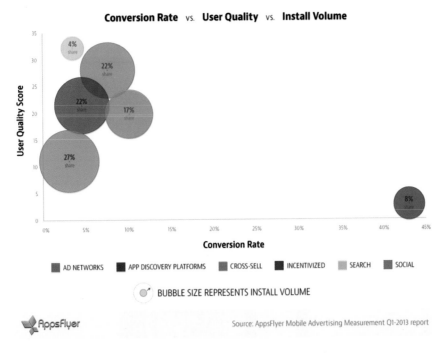

Conversion Rate vs. **User Quality** vs. **Install Volume**

Figure 11.11: The various channels that lead users to your app also factor into its success.

When I wanted to attach a company logo to some freelance work I did years ago, I thought about what I wanted my work to represent. My work up to that point had been redesigns of printing materials. In a sense, it was a "transformation" of previous work. Since my last name is Murman, I combined the two words into "Murmanations." Cheesy as it was, I felt the name described what I wanted my work to mean. The graphic I came up was the letter "m" in two different fonts combined into one.

Screen shots need to take the logo one step further. You must graphically show your app in action to give users a feeling for what they can expect. The trendy shots include holding a device and showing hands interacting with different screens, but it's not completely necessary to do so. In the end, put your best foot forward.

Utilize Your Users

There are other aspects to the app stores that can seem out of your control, but great app makers engage users by using ratings and reviews. As with your original user interface design, find intuitive ways to engage users by asking them to rate and review your app. This can be the most challenging aspect of putting your personal work out there because you have to take the good with the bad. There is always good to be found in comments, however.

If a user gets hung up on a screen and then gets frustrated, this can often lead to a negative review. What an invaluable piece of information! You can now incorporate this into an update that makes the signup process, navigation, or other commonly used feature easier to use.

Many developers get hung up on ratings, and for good reason. High ratings can get you listed on the home pages of app stores and noticed by review sites. If you have only a few thousand users all of whom leave five-star ratings, that can be more powerful than ten times the number of reviews with ratings across the board.

Figure 11.12: Facebook has seen huge growth in its mobile revenue because of an improved interface and advertising platform.

Marketing Is a Job for a Reason

Even though social media ranked much lower than app store searches in the previously mentioned Nielsen and Forrester surveys of how users find new apps, the Internet can be a big part of spreading the gospel of your development genius. Engaging your more prominent users such as bloggers and tech site editors might just be the tipping point you need for a jump in downloads.

As sad as it sounds, we in the tech-writer community are an incestuous bunch, and we all write what others have already posted. Many of the reviews I have written for new apps and hardware originated from a post I have seen somewhere else. Please don't make assumptions about this, because even if this displays a pack mentality it generally means we all like cool stuff and want to share our thoughts on it.

In the old days of advertising, there were only a few different publications where you needed to get an ad or a write-up. Now, with the exception of a few high-profile websites, it is near impossible to know which writer can get you the publicity you need. The key is to keep soliciting and sending your app out for previews.

KEEP UP TO DATE WITH APP STORE RANKING ALGORITHMS

As the app marketing blog Fiksu noted in August 2013, Apple has recently made some changes to the ranking algorithm that it uses. Some key factors that now come into play are:

- Ratings, as well as downloads, now affect rank.
- The positioning is updated every three hours now instead of every 15 minutes.

The goal appears to take more into consideration in determining rankings than merely number of downloads and five-star ratings. This can open the door for more quality engineering to take the place it needs. There are some things, however, that still need to be considered: download numbers and ratings still count, as well as quality. Just not as much as before.

When you are beta testing, it can sometimes be beneficial to reach out to some writers you follow online and gauge their interest in checking out a preview. While many of the more well-paid writers may get these sorts of e-mails daily, you might be surprised. As I mentioned earlier, tech writers like impressive designs that accomplish tasks in a new way.

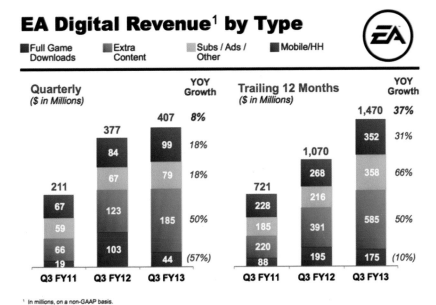

Figure 11.13: EA has taken full advantage of its mobile platform of games. Its largest area of growth is mobile, led by freemium gaming.

Also, don't forget about your already established users. If you are capturing their e-mail addresses (which you should always do), you have a ready-made e-mail list of users from whom you can gain feedback. I'm always impressed when I sign up for a new app and then get an e-mail from the company's founder asking for my thoughts on the software I recently downloaded. You can also engage with lapsed users who have moved on to find out what you could have done better.

All of this feedback is important, because even a negative review or e-mail does not mean your software is dead. With the increasingly short news cycles online, you can turn a frown into a smile sooner than before. Just be transparent with your software; then inspect and adapt.

Advertising Options

If there is a theme to this chapter, it centers on the idea of noise and how you can rise above it. That said, there are just times when you need some professional help to get the word out. All of the statistics quoted are evidence of how app marketing is turning into a cottage industry. Some marketers focus on ASO, some on capitalizing on ratings and reviews, and some on paid advertising. The best have some sort of analytics engine that includes paid advertising to help focus your work in the right direction.

Google makes banner advertising easier than ever if you want to incorporate a marketing budget into your development costs. The core business of the Internet giant continues to evolve and allow for broadcasting your message through paid content as a viable option for gaining downloads. While agencies can assist with this, you can easily do on your own as long as you have a focused set of branding materials to use.

Your app's website should be at the center of your brand. The tone and look need to match with the UI you have submitted. Matching banner ads help keep potential users focused on your message and make finding the app in the store easier.

Final Thoughts

Just as there are many successful pricing models, different apps have become successful in many ways. Nobody expected Twitter to emerge from South by Southwest in Austin with triple its previous traffic and to ignite a social media revolution with mobile devices. Its success helped make the spring arts festival in Texas the place where new brands want to emerge, but its enormous success is hard to replicate. Making your app free or available at a discounted price may help bring users to the store or just create another icon lost in the fray. Sending out beta releases to the mobile community can generate the kind of buzz an app like Mailbox received prior to release or go generally unnoticed.

None of this should discourage you from going ahead with your idea. It should just further emphasize the importance of that idea. The startup founders of the world have made a living delivering pitches for their ideas. Start your pitch with your friends and family; they are the proving ground you need. Before you write a line of code or sketch a logo or write a word of copy, focus on your idea.

Promoting Your Apps

The goal of developing an app is not the technology but the business it can generate. Business is generated only when customers are engaged. This chapter focuses on how to market your solutions and how to engage with your customers and drive a new wave of commerce through a smartphone or tablet.

There are dozens of books, hundreds of articles, thousands of blogs, and millions of tweets that offer information on how to be a better marketing person. This chapter does not give you all of the tools you need to be a marketing genius but outlines pragmatic tools my clients have used to become successful in promoting mobile apps.

To keep things simple, I want to focus on three concepts. They are:

1.   Mobility as a new communication channel to your customers

2.   Cost-effective marketing

3.   Cross-branding

The bottom line to my approach to promotion is this: everyone can compete with the major players. For every major powerhouse there are dozens of one- and two-person companies that have been successful in making lots of money with apps. Doodle Jump, Tiny Wings, and Clear Day are examples of apps that have generated thousands to millions of dollars for their developers.

Using Apps as a New Way to Drive Sales in Your Company

Do you have Google Analytics measuring who comes to your website? You will be surprised to see how many of your customers are now arriving on your site using a mobile device. The number has doubled every year for the last three years. By mid-2014 your customers will be using mobile devices to view your site more than any other type of device. Do not treat mobility as a fad.

Creating a mobile app requires that you treat the new digital media as a business channel. Start with the goal of building a new business, not a sideline. This will change how you execute your business.

DEVELOPING YOUR APP MARKETING PLAN

The job of the marketing department is not easy. The channels you need to support are increasing. Between traditional, social, and emerging technology you really have your hands full. The challenge is that you cannot dilute your message to any one channel. As humans we still watch TV. We still read magazines. We are still influenced by ads on the radio.

Your marketing plan must now connect multiple channels. Martha Stewart is a pioneer of cross-promotion. Her books promote her TV show, which in turn promotes her magazines. As a promoter, you must do the same for your apps.

WORKING WITH LOCAL PRESS TO BUILD AWARENESS

The global economy is fragile. There is no doubt about it. No matter where you live it is always a challenge to promote positive news. This is where you can step in as an entrepreneur. Contact your local press and ask it to run a story about your amazing, entrepreneurial company.

Every city wants to host the next Silicon Valley. Technology companies are a reflection of wealth. Who would not want that? As part of a technology company, you can leverage the local press to illustrate that technology is being developed and promoted locally.

When you pitch your app to local press, you want to highlight the jobs the app will bring into the local region, how it will promote the area, and how it can help people. In other words, keep the focus local. Getting promotion from local press is the first step to becoming a PR powerhouse.

BECOMING A PR POWERHOUSE

Are you Don Draper? You may not have the finances to hire the team from *Mad Men,* but there is no reason why you cannot become a media powerhouse. LinkedIn groups, Google+ Communities, and Facebook are great places to meet other PR leaders and pick their brains for ideas.

The first step to gaining credibility is to get published. You will find that media companies will want to listen to you when you drop the mythical words "I was published in *XYZ* magazine." Publication validates what you have to say.

The goal you need to set is to use each media event as a tool to leverage the next opportunity. You need to consider how a local newspaper can get you onto a local TV station, in turn leading you to presenting at a local event, leading you into larger press outlets. The opportunity today is that we all have many different ways of consuming content. Your audience consumes many different types of media, and you must be prepared to cross-promote each channel.

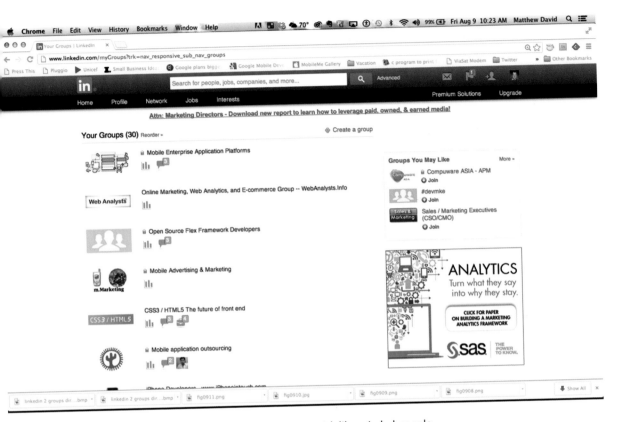

Figure 12.1: LinkedIn groups are a free and easy way to connect with like-minded people.

Can you skip all of this and hire a PR company? Perhaps. I have had mixed results with PR companies. Some have been amazing, and others merely sent out press releases. The end result I have always arrived at is this: the press really cares about whom they are interviewing. You need to be engaging, passionate, and driven. This is why I recommend that you engage with the press to begin with. Use the local press as a way to practice until you are presented to the bigger players.

CONTACT RADIO SHOWS

Radio is brilliant. I have a slight bias as I was brought up in a house where my mom was on the radio every week. It was common to have DJs milling around the house (and, believe me, they all had faces for radio).

What I did learn through this upbringing was this: radio is starving for content. Do not be afraid to drop in and talk to your local news radio show. Technology is always a hot-button topic, and your mobile app story will be a really popular subject.

The key to radio is remembering that the audience cannot see you. A great piece of advice I was given very early on is to describe everything you see when talking on radio. This does take a little practice, but it repays you back later when you are selling every detail of your product.

As well as traditional radio, you will also want to check out online podcast stations. There are loads, and I have found it really easy to get onto the shows. What I really like about podcasts is that you can access the station later and e-mail the show to a potential client.

PUBLIC SPEAKING

Are you the leader of your company? Then get comfortable speaking in public. There are many avenues for public speaking, but it essentially comes down to this: a group of people wants to hear what you say.

There are few people who are born public speakers. You will need to practice. Here are some tips I have been given to help with speaking:

- Thank the group that invited you to speak and for the audience coming.
- The audience is there to hear you—do not rely on a thousands PowerPoint slides.
- Use stories to illustrate a point—highlight what other companies or groups are doing that reinforces what you are saying.
- Really be an expert on the content—you will be asked questions, and you need to have quick response.
- If you do not have an answer, then admit to it.
- Jokes are good only if you can deliver comedy. If you cannot, then do not force a skill you do not have—don't worry, you have many more skills you can use.
- Prepare responses for questions that may come up—it is always good to keep some gems for the questions section.

Check out meetup.com to find the local groups you can meet and speak at. There will be all sorts of groups you will speak to. Some nights will be great, and the audience will be alive and listening to every word. Other times you will want to get out as fast as possible.

The reason you speak is simple. You are selling influence.

GUEST BLOGS

You can start your own blog and promote your own voice. That is a great place to start. The value of your voice increases when you are seen via different news sources. Writing guest blog posts for different groups is another great way to introduce your ideas to a new audience.

When you do write a guest article, remember to promote articles from your own blog. Google tracks how often you are quoted in the media, and having a lot of quotes will raise your standing in its relevance ranking.

GET REVIEWS FOR APPS YOU CREATE

At the end of the day, you can tell people how great you are. This is good. What makes you great is when people tell you how brilliant you are. That is your goal.

Submitting your apps for review gives you the opportunity to have an impartial group state how good your solution is. A review may be written and a grade given. Normally, there is something in the review you can grab and use in your other promotion efforts, such as "Best App of the Year"—Billy's App Review.

Reviews are a validation service the movie industry has been using for decades. As consumers, we see only the statement in the review. Very rarely do we read who gave the review.

Cost-Effective Marketing

Something magical is happening in marketing—new slots of time are now becoming available for marketing.

Social media are a clear example of a form of communication that is taking on an increased presence throughout the day. If you are like me, your phone is your alarm. When I turn off my alarm, the first check I complete is to see how many tweets I received during the night. During the day I find myself checking my phone when I am in the elevator, waiting for the kids, or standing in line to buy groceries.

Emerging technology is targeting pockets of time that we previously did not use. Smartphones make it easy to complete a quick check without the hassle of booting up a computer. You can also use this time to make a quick post to your social media group of choice.

WORKING SOCIAL MEDIA

Social media do matter. They are the only channel where you can connect directly with your customers, no matter where they are in the world. There is a good chance that you are already using social media to keep in contact with friends and family.

Using social media to promote your apps is not the same as keeping in contact with your family. Your app has a focus, and your engagement on social media is as an expert on the subject of your app.

Your obligation with social media is to become the face of your app, promoting the subject matter it is built around. Don't just sell. You will be burned very quickly online. Be honest about the subject matter and share your knowledge. Educating your followers makes them advocates of your ideas.

FACEBOOK, LINKEDIN, TWITTER, AND GOOGLE+ ALL MATTER

So which social media channel do you choose? How about all of the above? Your audience is fickle. Social media channels grow to millions of users very quickly. I am not encouraging you to join every startup that has "social media" in its company mission statement, but I do encourage you to keep a close eye on the established and growing companies. And, importantly, keep an eye on the social media companies that are losing followers (you do not want a campaign built around MySpace.com).

Figure 12.2: Use a gimmick to get attention on social media.

The leading social media channels today are Facebook, Twitter, Google+, and LinkedIn. Each provides different types of content to your audience:

- Facebook is social and deeply personal.
- Twitter offers news aggregation and "kiss in the dark" connections.
- Google+ is strong in building communities of like-minded people.
- LinkedIn is the professional network where you put your business face first.

Take each social media platform and "learn" how to engage with people using it. After a while, you will see trends on how to engage.

The next step is to promote content. This is where the content you develop on blogs and guest sites really pays out. Syndicate your ideas.

There are many tools you can use to help with your social media communication, such as:

- Content management—Hootsuite and Pluggio.com
- Managing followers—Manageflitter.com
- Influence Ranking—Klout, Kredd

The market for social media is new. You will need tools to grow your presence, and you should regularly review new tools to see how they can benefit you.

JOIN DEBATES, ASK QUESTIONS OF SPECIFIC PEOPLE, AND DON'T BE SHY

Social media give you something that was very hard to get until recently: access. It is very easy to access anyone. Last year I was working with a startup and was following several writers on Twitter. One said he was bored and needed a good story. In 140 characters I pitched our product.

Later that day, after a follow-up phone call, our company was on the front page of Fast Company. The article was so popular that it stayed there for two weeks and brought more than 250,000 unique visitors to our app. All of this came from reaching out to a journalist on Twitter.

Oh, and the journalist lives in Portugal, and I live in Wisconsin, and we had never spoken to each other before.

Don't be shy. Reach out to people and engage. You will be amazed at the results.

Cross-Promoting Your Apps

Your apps will be popular. Unlike websites, an app is a captive product. Websites are plagued by the reality that your audience is just one click from leaving your site. There is no "one-click" loss in an app. Someone has chosen to install your app. Every time your app is opened, you have an opportunity to sell to your customer.

ADVERTISE YOUR OWN APPS IN YOUR APPS

There are three types of apps: free, paid, and semifree.

Free apps provide their content for free. These apps often rely on advertising to support future development.

Paid apps, such as Angry Birds, generate a burst of revenue each time a customer buys an app.

Semifree apps are apps you download free of charge but that allow you to buy upgrades in the app through an "in-app purchase." Temple Run receives the majority of its profits from in-app purchases.

For all of these three models you need to assess how you use the captive time with your audience. For free apps you must use advertising to keep profits coming in.

Paid and semifree apps are different. The revenue is generated by the sale of the app or a feature in the app. For this reason, use space in you apps to cross-promote your products. A great example of this is Angry Birds.

THE ANGRY BIRDS APPROACH TO ADVERTISING, MERCHANDISING, AND SELLING

In many ways, Angry Birds is the definitive app for mobile. The game is easy to learn, uses touch-screen features unique to a phone, and can be played for a little as a minute or for hours.

The success of Angry Birds cannot be diminished. More people play Angry Birds than any other game in history. Any game! Every phone, tablet, and game system has Angry Birds installed. In a world marching toward billions and billions of devices, we are talking about billions of installations of Angry Birds.

Billions is a big number with incredible reach.

The owners of Angry Birds cross-promote their own products in their apps. At the end of each level there is an ad for other Angry Bird games (Rio, Space, Star Wars, Seasons, Piggies, and the new Star Wars Angry Birds II), but you will also see ads for Angry Bird merchandise, cartoons, and news. The goal is to keep you in the Angry Birds universe of products.

The idea is not new. Disney pioneered cross-promotion years ago with Disney movies, theme parks, and merchandise. The first Angry Birds full-length movie is coming out in 2018.

DON'T FORGET BILLBOARDS, TV, AND TRADITIONAL ADVERTISING

At some point, when you are generating enough revenue from your apps, you will want to consider expanding your marketing efforts. Do not forget traditional methods such as billboards and TV. Both are highly effective.

Our world is migrating to a world of commuters. We spend on average 90 minutes behind the wheel every day as we go to and from work. Billboards are still a key part of all of Apple's advertising. The "Think Different" and "iPod" billboards won awards. Billboards are inexpensive. Use them to catch the attention of your audience.

While the naysayers are telling the world that we are watching less TV, the advertising money spent on TV commercials is telling a different story. We like TV shows. We want to be entertained. Traditional broadcast TV is changing, but the viewing figures for cable and Internet shows are exploding. With TV you have a captive, often-social audience. We like to watch TV together in groups. Leveraging our awareness that the audience uses smart devices while watching TV shows gives you the opportunity deliver a second screen promotion. The makers of Oreo cookies performed this trick brilliantly during the 2013 Super Bowl. Oreo was a sponsor of the event and ran a commercial with a well-oiled social media team in place to interact with followers on Twitter and Facebook. Then the world went dark as the power failed at the Super Bowl. The Oreo team responded with an ad stating "you can still dunk in the dark" that was retweeted thousands of times: old media and new media working together.

The bottom line for promotion is to be active. Do not sit back and assume your app will start selling to the tune of millions of dollars. That simply won't happen. However, with a little effort, you can build out a successful media campaign without having to spend millions of dollars. Promotion can seem daunting at first until you realize that you have a great story to sell and that the media loves a new and positive story. In fact, you may find you actually come to really enjoy promoting your apps, your team, and the way you are changing the world. Every company started as an idea and became something much more. Success is something you can convert into free promotional dollars. Become the evangelist for your product.

CHAPTER 13

Future-Proofing Your Apps— It's Going to Be a Bumpy Ride

When I first became a father, the best advice I was given relates to the constant changing world of technology: just as soon as you figure something out, get ready because it's about to change. With each release of Android and iOS, new and inventive methods of allowing users to interact with their software are established. It was once enough for iPhone users to just touch instead of clicking; touch gestures were introduced. Of course, anyone who has used a Galaxy S4 or Xbox Kinect knows you no longer need to touch a device to interact with it.

This affects developers in various methods. Perhaps your billion-dollar idea was not possible before new technology or regulation allowed it. Your app may not gain traction on smartphones or tablets. It may not work on an existing platform, period. The reasons for continued iteration upon your development go on and on.

The purpose of this chapter is to give only a few examples of how the changing landscape of mobile technology will affect the future of your work. It is not a complete list, just something to get you started.

How Can You Iterate upon What You Just Built?

Even with the number of apps available in Apple's and Google's respective app stores, it is no easy feat to become one of them. After you successfully submit your app, it's even harder to become a success. Even success itself can be fleeting once you achieve it. You must find a way to continue to stay relevant to your users, with more releases to your current offering or a new offering altogether.

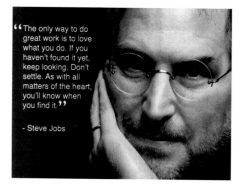

"The only way to do great work is to love what you do. If you haven't found it yet, keep looking. Don't settle. As with all matters of the heart, you'll know when you find it."

- Steve Jobs

Figure 13.1: Steve Jobs pioneered the need for a standard of excellence in technology.

You might be asking yourself how you can catch lightning in a bottle again.

What separates thought leaders from the rest of us is the ability to see what is just under the surface and apply it to their area of expertise. Software engineers have been known to reference Beethoven's Fifth Symphony to describe perfect architecture. A great sculptor can reference the genius of Edison's light bulb design. For inspiration for your app, however, look no further than the great Steve Jobs.

In a blog post for Medium.com, the writer Kevin Ashton analyzed one of the best questions the former head of Apple asked on a regular basis:

"Why doesn't it work?":

"Sales + Customers = Nothing Broken" is the formula for corporate cyanide. Most big companies that die kill themselves drinking it. Complacency is an enemy. 'If it ain't broke don't fix it' is an impossible idiom. No matter what the sales, no matter what the customer satisfaction, there is always something to fix. (Retrieved from http://wolfgangwopperer. tumblr.com/post/47180309787/sales-customers-nothing-broken-is-the-formula)

Of course, many have written about the former head of Apple and his attention to detail. We didn't need to be informed of that. Articles, books, and movies have been telling aspiring creators for years how awesome Jobs was. What stands out is how a simple question about something like the current viability of a product can change your view of it.

Today's Standard	Emerging Capabilities	Future Focus
Transaction Engines	Interactive Devices	'Smart' Mobile
• Text banking • Check balance • Pay bills • Money movement • Locate ATM/ Branch • Mobile RDC • Help content	• Geolocation (e.g., merchant-funded rewards) • Targeted marketing • Social Media Integration • Actionable Alerts • PFM tools • Cardless Cash Withdrawal • Streamlined P2P	• Gamification • Remote Bill Capture • Voice recognition • Various uses of camera / imaging • "Just in time" services (e.g., budgeting, finance)

Figure 13.2: Keep an eye on the present as well as the future to push your software to the next level.

Jobs was famous for asking this question about all products, including his. There is always something that can be refined to make something better. Often, this contradicts the stance companies have regarding their offerings as "good enough."

Not that they would ever admit that. They just don't work on improving. I once worked for a company that was an industry leader several years ago. That can last, however, for only so long before competitors catch up and start to put pressure on you. While we still provided our customers an amazing suite of products, I am not talking out of school too much to admit that we did not maintain our lead in some areas.

The lesson here is that catching up takes up a lot of energy. Regardless of the industry, most companies know what that feels like. Many who don't have a mobile strategy in place should be feeling this strain right now. When looking at your product line, don't think of the things you like about it right now. That's for the marketing department to discuss. Product people need to look at what's wrong. Ashton's equation of sales plus customers equaling nothing broken is really dangerous. You may have customers now, but your competitors are selling currently as well. Nothing being broken can turn into broke really fast.

PIVOT FAST AND DON'T BE LEFT BEHIND

Just as fast as they entered into the mainstream, QR Codes started fading in popularity when devices didn't embrace the technology. Apple is attempting to do the same thing with NFC chips, trying to make BLE the new standard in wireless sensor communication. It's still too early to tell if responsive Web design is the answer everyone expects it to be.

The point is that you can't put all your eggs in one basket, expecting it to keep filling up. There will come a time when your feature or enhancement will be replaced by something shinier to users. At that moment, you can't be afraid to cut ties and move on. Only then can you be expected to keep up with the ever-changing world of the technology we love using.

Instead of waiting for that to happen, look for something to break on your own. Perhaps your platform needs to be rewritten, but to do that means breaking it down and slowly building it back up. Instead of being upset, your customers will applaud your desire to improve their experience or to add features more easily. It takes more energy to catch up after you have let someone else help you realize that your product is broken than if you had realized it on your own. Go find something and break it.

That said . . . don't overdo it.

Imagine a new restaurant opens up down the street, so you take your family out one evening to check it out. After being seated, you take a gander at the menu and notice it is filled to the brim with items. Double-digit numbers of items in all the categories: appetizers, salads,

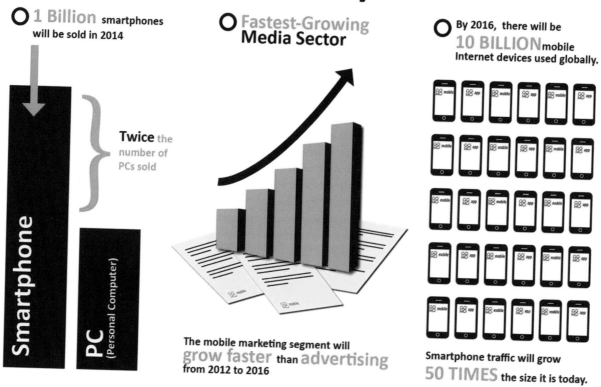

Media Analysis

○ **1 Billion** smartphones will be sold in 2014

○ **Fastest-Growing Media Sector**

○ By 2016, there will be **10 BILLION** mobile Internet devices used globally.

Smartphone

PC (Personal Computer)

} **Twice** the number of PCs sold

The mobile marketing segment will **grow faster** than advertising from 2012 to 2016

Smartphone traffic will grow **50 TIMES** the size it is today.

Figure 13.3: Smartphones have replaced the personal computer as the most commonly sold piece of technology.

entrees, desserts, you name it. You can't make your mind up, and even if you do there's a sinking feeling that you probably made the wrong decision. Too many choices diminish your experience.

Product decisions are often made based upon concepts such as cost and value instead of single features people want to use. At the risk of upsetting my architect buddies, I run into this issue many times when talking to them. Complexity is not necessarily the hallmark of well-designed software. It's nice if you are designing a telecommunications billing system, but not if you are trying to make our mobile devices more valuable.

Dropbox is a product that focuses on one feature (the feature is very robust, but you get my drift), and its designers make it great. They have amassed more than 100 million users because they made cloud storage easy and secure.

Not every product can be successful with just one great feature. As a result, the rest of us have to figure out the right number of smaller features to package into the right compilation to make it attractive to the customer. That's when stakeholders often make a mistake. Maybe the development framework to release multiple applications is really expensive, so they must find a way to cram as much into a product (and release) as they can. The first release is the most important, or so they are told, so we must develop until our fingers bleed!

It's just not true.

If you have an idea that works, you can have lightweight releases focused on one or two great features that provide something your users want. Here's the most important part of this chapter. If your idea doesn't sell, that's OK. Start with another idea, make that feature amazing, and see how your users feel about it.

Failing fast, as they say, is much better than spending a ton of time to making a lot of features that nobody is going to use. Take a hard look at your road map. Are you trying to cram too much into a release? Have you thought through the value and necessity of each feature? Finally, is it all truly vital to have them all together? Asking those questions will show your team value in the decisions you make and your releases smoother.

Remember Your Canvas

When I need to be reminded how hard designing for the Web can be, I go to Yahoo's home page. Doesn't matter the browser or device you use; it's a mess. What the exercise does is provide a reminder of The Bubble. At the height of its power, the desktop Internet became a gigantic, bloated heap of graphical data thrown at you. Something was needed to disrupt designers who were beset with power. The biggest challenge many designers face today is how to balance all the customer requests on pages and keep things as clean as possible. Designers used to think, "What's one more button? I have all this real estate to use!"

Figure 13.4: Mobile payment technology is attempting to replace plastic with the digital wallet.

Now, with only a few hundred pixels of space, designers must decide what features actually matter. Instead of being "just one more feature," each iteration must be dedicated to ensuring a seamless experience for the user. That was never a consideration a decade ago. The same concept applies to mobile apps, because the canvas is the same.

So how do you decide what is important? Regardless of the interface, you can't make decisions solely on the basis of either data or educated guesses.

A designer I once worked with told me that it is easier to design from the phone up. If you can take what matters most and factor in the user's experience, tablet and desktop designs will follow quickly. Imagine if you took your platform and stripped the unnecessary away and then added as you had space.

What if you have to go in reverse? The task of trimming the fat from your app is as difficult a task as you can find as a product manager. Your users might react negatively to too much change. Even if the experience of the old version of the app was poor, there are users who will be used to it.

DON'T BE AFRAID TO REPLACE WHAT YOU TRIM

When the iPhone and iPad were first released, nobody imagined a day when they would be used side by side with televisions to enrich the viewing experience. The second-screen revolution now has movie studios like Disney including this enhancement with theatrical releases.

Parents were encouraged to bring their iPads to the re-release of the 1989 classic *The Little Mermaid* to have extra features unlocked during viewing. What was once taboo is now being embraced.

This goes to the point of expanding where you see gaps in fulfillment. Someone will be the first to take a chance on your new feature, so don't be afraid to find a way to use it!

When you look at the current UI at LinkedIn, you will not see a redesign that was sprung on users at once. The changes were subtle, making minimal changes that users had to get used to. Before I knew it, the old interface was completely gone, but it was replaced by something I was really proud to use.

To accomplish this feat, your road map must be well informed and planned. That doesn't mean you must execute it flawlessly and without deviation for it to be successful. That's where the Lean and Agile methodologies come in to play: you fix a little, and then you test a little. Act, learn, then build upon what you have learned.

If a change doesn't work, there's no harm in reverting to the old version or changing course. The goal is to make informed decisions, execute them, and then measure their efficacy. A masterpiece is painted slowly, even on a canvas 640 pixels wide and 1136 pixels high.

The Great Talking Disorder

Talking with users is the lifeblood of people who create great products. It needs to happen before design, during development, and after release. One of my favorite parts of my work is talking to people and companies of all kinds about what they do—not the product they create or service they perform but what they do to accomplish those goals. In essence, I get to ask the question "how."

Figure 13.5: Despite 4G and LTE technology, the mobile Web continues to disappoint and confuse users. This opens the door for native app developers.

Nine times out of ten, the conversation heads in the direction of connectivity. We desire better or more effective collaboration. Sellers collaborate with buyers, managers work more closely with employees, and peers share the workload with one another. Often, the notion of better tools is the answer. While I think it is possible to come up with fun ways to better connect our work days, I don't think this problem will be really solved with a better instant-messaging feature.

What we suffer from is called The Great Talking Disorder. Notice I did not use the word "communication." We do that through social media, texts, and e-mail a ton. What I'm referring to is sitting in casual or corporate settings and looking at one another while our lips are moving; preferably with sounds coming out.

Figure 13.6: Can your app communicate better than your competitor's?

Your users won't always be available for this type of collaboration, so look for inventive ways to achieve the same results. The feedback loop can be initiated and completed in a variety of ways, especially with the technology surrounding mobile devices.

Think Beyond Today's Devices

One of the greatest parts of my job is to read report after report from analysts on current and future trends. For the most part, you have to take a lot of research with a grain of salt. Trends are in the eye of the beholder, just like in the stock market. One of the ideas I have been reading about over the past few months is that now is the time to launch your own platform. Of course, it's impossible to do that without a killer app.

Remember the days when our mobile phones kept getting smaller and thinner? Apple still thinks it's an important feature. I am starting to think that our mobile phone needs just to be small enough to fit in our pockets so that it can be connected to my wearable devices.

The days when we were scared about the data our devices gathered are basically over. The more I can learn what is happening with my body, the weather, the food I eat, and countless other data points, the better. Add in messaging, GPS, social media, and call alerts, and the platform is complete.

Applications may not be enough to compete in the mobile sphere alone anymore. With app stores still in their infancy, that is a scary idea to float out there. I will be keeping track of the wearable device market with great interest this year. If it's not already a reality, 2014 may be the year of the wearable platform.

Figure 13.7: The Samsung Gear is attempting to pioneer wearable technology much as the iPhone and iPad did for mobile devices.

Smartwatches are not the only avenue to watch, though. A quick pass through Kickstarter's projects show a multitude of small devices that can gather and enhance our mobile lives. They all have similar features: they're all small, cheap, connected products that allow mobile technology to report. Let me tell you about one that is ready to disrupt our devices.

As a fully funded project, Tile is now making an affordable way of tagging individual items and linking them to a cloud-based system. That means you can use your—or anyone else's— mobile device to locate these items. There is even a function that makes the square tile (which looks to be a couple of inches in size tops) beep for easy location.

By far the greatest feature is the system each tile is connected to. Imagine you've tagged your bike, and it gets stolen. You can mark the item as lost, alerting the system to start using every mobile device connected to the app to start looking for it. An unsuspecting Good Samaritan could help you locate your lost item without even knowing it.

At the current price of $18.95 apiece, buying the max of ten tags gets a little pricey. However, spending $100 for five items can get my family's two cars, two sets of keys, and my wife's purse all tagged. I can now park with confidence out in public.

When the company takes off a bit, I'm sure devices will get cheaper with longer battery life (each tile is reported to last a year). I'm not sure what the tipping point will be on price, but this is a great start.

Which Device Is Your App Right For?

The past year or two have been banner years for mobile development, if only because they have offered the first real opportunity to gather data on how people use their mobile devices and how marketers can properly engage them. Many in the biz know that desktop users are classified in the "research" stage, highest in the funnel (meaning farthest away from a purchase). They are gathering information, mostly to bookmark or come back to for further investigation.

Tablet users, once thought to be akin to mobile phone users, are almost as high in the funnel as desktop users. Nearly three-fourths of tablet users engage their device in the home. While there is a little bit of difference between these two groups, there's not much. It makes sense to make the tablet version of your site resemble the desktop version since they will be used in similar ways.

Phone users, however, are much further down the funnel. Users search and seek on phones when they are out and about, ready to make a purchase. I found myself in this particular situation recently when I found that the walk-in clinic I used my desktop machine to locate was not for walk-in customers. I had to then use my phone to locate the closest acute-care clinic.

Figure 13.8: Knowing the right device for your app can aid in its success before you begin a single day of development.

Thanks to a mobile case study, Buffalo Wild Wings realized that mobile device users were its target demographic for NFL Sunday promotions. Done in conjunction with Nielsen, the study found that exclusive smartphone advertising led to an increase of purchase intent for 45 percent of users. BWW noticed that customers would check scores on their phones during games and that the best way to reach them to promote specials was on that screen. The most telling remark was from the company's director of digital media, Ryan Richardson: "It showed us using apples-to-apples measurement that we were getting people to say that they would go and fewer to say that they don't go to us through delivering this mobile experience," he said. "Though it's new, I answer to metrics. The only way I make money is when people come to my restaurants and buy wings and beer. It's really cool to deliver an awesome experience, but if that's not driving them into my restaurant door, it's not making me money."

While it is not the case for every product category, there are mobile application for many. The campaign can be customized for the place in the sales cycle where your users reside. If you can catch them right before they are ready to purchase, you will see a huge return on your marketing investment.

DON'T OVERLOOK WHAT'S RIGHT IN FRONT OF YOU

Often, we get so concerned with making a product "right" that we overlook the obvious. A blog post from UX consultant Harry Brignull perfectly illustrates just how unnecessary some product decisions are. He says, "If you tell it to work out 200 factorial minus 200 factorial, it will do a lot of unnecessary computation, and perhaps produce an overflow error. The intelligent solution is a far more lazy one."

That means the many screens describing what your app does may be seen by your users as a waste of time. They can figure it out on their own. Instead of utilizing an account creation module with four or five screens, just add a Facebook login button. The data you get will be the same, and users will be happy they didn't waste a few minutes on an app they might not like in a few weeks.

Perception Is Everything

When people click the "like" button in Facebook's app (or mobile site), they assume the action is immediate. There are many times you like something and by the time you pull up the desktop site on your machine it most likely appears. Seems like there is no problem, is there?

As most in the tech community know, the truth is far different.

I don't doubt that Facebook has performance on its flagship product—your news feed— humming like a sports car. A "like" is a simple action, and, depending on what kind of network connection you have, there might not be any time lag. On a more complex action, a definite lag occurs.

When you create an account in a mobile app, for instance, the activity is a little more complex. There's information to write to a database, device information (e.g., location, contacts) to import, settings to save, and so on. The user does not see all of that. All that matters is that whatever you entered is immediately reflected once the new page loads. To do this, apps use the local database on the device to aid in performance while the actual saving of that data to the Web service takes place.

Let's go back to our example of "liking" something on Facebook. Even though the activity does not happen automatically every time, the app itself shows the action happening instantly. Users can rest assured that once the button is highlighted blue on their feed, they can move on. Can you imagine what would happen to the user's experience if there were a spinning wheel every time a stall occurred in performance?

In a 2013 Compuware survey, 48 percent of users said they would be less likely to use an app again if they were dissatisfied with its

Figure 13.9: Facebook's app makes you think things are happening fast, right?

performance. Half of your potential users will disappear if it runs slower than they think it should. I don't remember every post I have liked. I just know that I have liked a ton of stuff and I want the app to run smoothly while I do so. If not every action is saved for future reference, no big deal. My experience while in the app is what brings me back for more.

If you can resolve any latency issues, your users will be more forgiving of back-end data-storage activity. Of course, if you are developing a banking app, your priorities will shift. For most consumer applications, just make sure your app is perceived well. That's really the only reality you need to increase usage and engagement.

OPTIMIZE IN EVERY WAY

When iOS7 was first announced, there were those who both applauded and jeered the new design. What hasn't been up for debate, however, is the new library of code optimizations. Developers can more seriously develop apps with battery conservation, improved performance, and additional gesture navigation.

Yes, designers had a field day proposing how they would have done things differently. Apps would need to be reworked to include the upgrades users will now expect to see. Don't forget, though, to make every performance optimization you can with the tools Apple and Google provide. They do their work with you in mind.

Beware of Future Regulation

If some of you are still on the fence regarding the coming mobile revolution, take a gander at this stat: by 2017, the market research firm Research2Guidance estimates, the mobile health market will be worth $26 billion. With that comes a certain level of government regulation.

As a type 1 diabetic, I have many opinions on the FDA and its death-like grip on health care in this country. Should I be grateful that the government is trying to keep standards of care high in this country? Black-and-white answers tell only part of the story. The truth lies in the middle, with money being the driving factor behind most legislation today.

I read a report recently that ranks the United States seventh in the world in terms of quality of health care

Figure 13.10: The more mobile software becomes a part of our lives, the more Big Brother is going to have a say in how it should work.

provided. So it would seem that there is a paradox. I'm not merely stating the FDA is running health care much the same way the mob provides "protection" to local businesses, but some people might see it that way. Treating diseases is a much more profitable business than curing diseases.

I would argue that the coming mobile revolution in health care could be the tipping point in finally giving patients access to their own health information. I would not presume to know what to do with that information better than my health care provider, but every time my doctor entrusts me with my own data points it results in my taking better care of myself. My hope is that the government will agree with that point in deciding how to regulate this booming market.

There is more to regulation than this solitary point, which of course is what makes politics such a fickle beast. There is abuse to be had all around, mainly because there is a ton of money to be made. All of this research costs money, and there will be many a failed startup in this industry. In the end, make sure you vote on applications that can serve the greater good in a positive way. There are three easy ways to do so: do your own research, let your congressional representative know, and vote appropriately with your dollars.

In the End, Act First

The world of technology continues to change at an ever-increasing pace. There will be different technologies you will use in the future such as in-car systems, wearable devices, and smart TVs. While you can research them all, the solution your product provides must be in tune with the people it serves. It's true that capturing every point of data possible about the people who will interact with your code can be the key to delivering the kind of software needed to sate passionate audiences such as mobile device users. Many engineers struggle, however, to know when enough research and data are enough let them to move forward with a project.

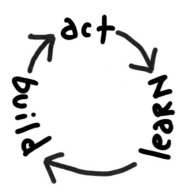

Figure 13.11: Planning is overrated. Do something about your idea; your competitors aren't worried about learning until after they release.

I spoke earlier in this chapter about your road map. Don't let the concept of planning out every release frighten you away from the first, which is the most important.

Traditional software development started with planning. Anyone born before the 1980s knows the pain of a six- to twelve-month release using the Waterfall framework. That's where Agile came from: a generation of frustrated engineers who had been beaten into submission by a sea of planning paperwork and requirement-gathering meetings. That's where the acronym "PDCA" came from:

Plan your strategy—This often took just as long as the development stage.

Do—Build and test your potential solution. Notice the word "potential," because all that effort was often for naught.

Check—You read that right. We didn't get to release the software until we were sure all the planning and building led to a workable product.

Act—Finally we got to start learning how the product functioned in the wild with real users.

As incredible as it sounds, nearly every software company in the 1970s and 1980s made mountains of money with this strategy. Your users won't wait that long, though. In today's fast-paced industry, where new devices are being released on an almost weekly basis, you must put your ideas out there fast to see if they can swim. That's where Eric Reis and the Lean Agile community came to the rescue. The old acronym was replaced by the more modern and agile ALB:

Act now—Put your idea out there. It doesn't take that long to get a piece of software written today for mobile devices.

Learn immediately—Start gathering information right away to see how your product is received.

Build upon it right away—You can do better than that. Start iterating right away.

Thus, the Minimum Viable Product (MVP) was born. While the term has been around long enough to start generating naysayers, it is hard to argue with its efficacy. It is past time to start admitting our work isn't always done well the first time. We will fail. Even if your code is up to snuff, you won't always get every feature or enhancement done in time for the release. Failure is not a bad thing when you can just learn from your mistakes and build upon them.

Figure 13.12: Lean, mean developers follow the MVP strategy to get their designs into production.

When you deliver software in smaller segments for mobile devices, this concept becomes even clearer. The best apps do one or two things well. If you're really good, you will present users with an easy and pleasant way of completing them as well.

If there is one theme to take away from this book, it is to take heart and act. The skills you are learning while building your app make you more attractive for future employers; in addition, you might have the next great idea. You won't know for sure until you take one of the ideas you have and act on it. If it works, find out how you could have made things just a little easier for users, and they will start sharing with their friends. If it doesn't work, there are data to be gained from failure that will help your next UI design.

Success and failure are relative. Forward momentum can be gained from both, which should always be your ultimate goal in building the next great download on devices.

KNOW THE ELEMENTS OF LEAN UX DESIGN

There are variable levels of lean design and development. Just how lean you want to be depends on your app's goal, time, and resources available. When evaluating this framework, keep in mind these points:

- Your project must have a mission statement. This not only declares your intentions but helps define for the customers how success will be achieved.

- At the heart of your mission statement is a core set of executables that needs to be laid out easily for the team to break up into development iterations.

- To execute, you must produce and test your hypothesis many times. The sooner you can get an element in front of a user to validate the concept, the better.

- Feedback loops are paramount to success. Gather feedback via several methods to collect a variety of data you can use to learn.

- Don't be afraid to pivot as soon as you see that an idea isn't working. Fail fast. The question "how do I know it isn't working?" is asked all the time. Your users will tell you.

Index

Page numbers in italics indicate images.